EXPLORATIONS

Mathematics for Elementary School Teachers

FIFTH EDITION

EXPLORATIONS

Mathematics for Elementary School Teachers

TOM BASSAREAR

Keene State College

BROOKS/COLE
CENGAGE Learning™

Australia • Brazil • Japan • Korea • Mexico • Singapore • Spain • United Kingdom • United States

Explorations, Mathematics for Elementary School Teachers, **Fifth Edition**
Tom Bassarear

Editor: Marc Bove

Developmental Editor: Stefanie Beeck

Assistant Editor: Shaun Williams

Editorial Assistant: Zachary Crockett

Media Editor: Guanglei Zhang

Marketing Manager: Ashley Pickering

Marketing Coordinator: Michael Ledesma

Marketing Communications Manager:
    Mary Anne Payumo

Content Project Manager: Cheryll Linthicum

Art Director: Vernon T. Boes

Print Buyer: Betsy Donaghey

Rights Acquisitions Specialist:
    Thomas McDonough

Production Service: Graphic World Inc.

Text Designer: Terri Wright

Photo Researcher: Bill Smith Group

Text Researcher: Sarah D'Stair

Copy Editor: Graphic World Inc.

Illustrator: Graphic World Inc.

Cover Designer: Terri Wright

Cover Image: © Simon Brown/Alamy

Compositor: Graphic World Inc.

For product information and technology assistance, contact us at **Cengage Learning Customer & Sales Support, 1-800-354-9706**

For permission to use material from this text or product, submit all requests online at **www.cengage.com/permissions**
Further permissions questions can be e-mailed to
permissionrequest@cengage.com

Library of Congress Control Number: 2010937591

ISBN-13: 978-0-8400-6245-1

ISBN-10: 0-8400-6245-1

Brooks/Cole
20 Davis Drive
Belmont, CA 94002-3098
USA

Cengage Learning is a leading provider of customized learning solutions with office locations around the globe, including Singapore, the United Kingdom, Australia, Mexico, Brazil, and Japan. Locate your local office at **www.cengage.com/global**

Cengage Learning products are represented in Canada by Nelson Education, Ltd.

To learn more about Brooks/Cole, visit **www.cengage.com/brookscole**

Purchase any of our products at your local college store or at our preferred online store **www.cengagebrain.com**

Printed in the United States of America
1 2 3 4 5 6 7 14 13 12 11 10

CONTENTS

4   Number Theory   81

5   Extending the Number System   97

6   Proportional Reasoning   139

7   Uncertainty: Data and Chance   153

8 Geometry as Shape   185

9 Geometry as Transforming Shapes   241

10   Geometry as Measurement   303

Endnotes   E-1

Index   I-1

Cutouts

BASE TEN GRAPH PAPER

OTHER BASE GRAPH PAPER

OTHER BASE GRAPH PAPER

OTHER BASE GRAPH PAPER

OTHER BASE GRAPH PAPER

GEOBOARD DOT PAPER

ISOMETRIC DOT PAPER

POLYOMINO GRID PAPER

POLYOMINO GRID PAPER

POLYOMINO GRID PAPER

POLYOMINO GRID PAPER

TANGRAM TEMPLATE

REGULAR POLYGONS

EXPLORING THE AREA OF A CIRCLE

PREFACE

Having a textbook contain two different volumes created the dilemma of where to put the preface. We resolved this dilemma by having two prefaces. This one, to the *Explorations,* explains how the two volumes work with each other. The preface to the text more specifically describes the goals of the course and the textbook features.

I have been teaching the Mathematics for Elementary School Teachers course for over twenty years, and in that time have learned as much from students as they have learned from me. The *Explorations* and the text reflect the most important things we have taught each other: that building an understanding of mathematics is an active, exploratory process and, ultimately, a rewarding one. Many students and instructors have told me that this is the most readable and interesting mathematics textbook they have ever read. This is exciting to hear, because I thoroughly enjoy mathematics and hope others will too. At the same time, I know that far from loving mathematics, many people are actually afraid of it. If this book is successful, you will come to believe that math can be enjoyable and interesting (if you don't already feel that way); that mathematics is more than just numbers and formulas and is an important part of the curriculum; and that mathematical thinking can be done by "regular" people.

What Is It We Want You to Learn?

At its heart, the purpose of this course is to revisit the content of pre-K–8 mathematics so that you can build an integrated understanding of mathematical concepts and procedures. From one perspective, it is useful to talk about content knowledge (e.g., different meanings of subtraction, why we need a common denominator when adding fractions, and classification of geometric figures) and process knowledge (e.g., problem solving and being able to communicate mathematical ideas and solutions). Let me give some examples of the kinds of content knowledge you can expect to learn in this book.

Content Knowledge   Virtually all adults can do the division problem at the right. If asked to describe the procedure, most descriptions would sound something like this: "4 'gazinta' (goes into) 9 two times, put the 2 above the 9, 4 times 2 is 8, 9 − 8 is 1, bring down the 2, 4 gazinta 12 three times with no remainder, put the 3 above the 2, the answer is 23." However, few adults can explain mathematically what "goes into" and "bring down" mean, other than by saying "that's how you do it." However, once you realize that one of several meanings of division is that an amount is to be distributed equally into groups and once you realize that 92 can be represented as 9 tens and 2 ones, then you can explain why long division works. Being able to explain *why* gives one what we call mathematical power, which means that you can apply this knowledge to solve problems that are not just like the ones in the book. You can solve "real-life" problems, which is one of the most important goals of school mathematics.

$$\begin{array}{r} 23 \\ 4\overline{)92} \\ \underline{8} \\ 12 \\ \underline{12} \end{array}$$

Similarly, most elementary teachers recall that an approximation of π (pi) is 3.14, and they may remember some formulas: $C = \pi d$, $C = 2\pi r$, $A = \pi r^2$, but they can't explain what π means or why these formulas work. Developing a conceptual understanding of π is not as esoteric as many people think. Let me illustrate. Look at the circle at the right and answer the following question: Imagine we had several flexible rulers that were the same length as the diameter. If you wrapped those rulers around the circle, how many rulers would it take to wrap around the entire circle? If you actually do this, you find that it will take a little more than three rulers. Thus, one conception of π is that the length of the circumference of a circle is always a bit more than 3 times the length of the diameter. With this conceptual understanding, the formula $C = \pi d$ "makes sense." Ensuring that math makes sense is another key goal of mathematics education.

I have just outlined two of many explorations and investigations that you will do over the course of this book. What so many of my students have discovered is that if they have a chance to work with mathematical concepts in an active, exploratory manner, they can make sense of elementary mathematics, which means that they will be much more effective teachers. Much of the impetus for teaching to foster this style of learning comes from the National Council of Teachers of Mathematics (NCTM), which has published three sets of Standards in the past twenty years and ushered in a new reform movement in mathematics education. I find that many students who enter the course with negative feelings toward mathematics view the NCTM Standards very positively. Many tell me, "I wish this is what my mathematics courses had been like."

NCTM   The first NCTM Standards document, *Curriculum and Evaluation Standards,* was published in 1989. It sets forth a vision of why mathematics is important for all citizens to know and describes the mathematical knowledge one should develop by the end of high school. Chapter 1 of the text will explore the Curriculum Standards in more detail. The *Professional Standards for Teaching Mathematics* will be discussed later, and you should expect to examine the *Assessment Standards for School Mathematics* in your methods course. Since one of my goals is that you will use NCTM to frame your own teaching, I will discuss my vision of using NCTM Standards as much as possible. At appropriate times in the text, I will cite passages from the NCTM documents.

In 2000, the NCTM finalized a new document called *Principles and Standards for School Mathematics,* which represents an update and refinement of the 1989 document. A summary of the ten standards can be found in Appendix A of the main text, and the full document can be found at the NCTM website: nctm.org.

What Does It Mean to Learn?

As you will discover in Chapter 1, your attitudes and beliefs have a lot to do with how you learn mathematics. My beliefs about what it means to "know" mathematics have led me to create a very different kind of textbook, and it is important for me to describe my sense of what it means to learn. Let me contrast some traditional beliefs about mathematical learning (which I disagree with) with this book's approach and then describe how this book is structured.

Active vs. Passive Understanding   Many students believe that mathematical understanding is either-or (e.g., you either know fractions or you don't). In actuality, mathematical understanding is very much like other kinds of knowledge—you "sort of" know some things, you know other things "pretty well," and you know some things very well. The belief in either-or leads many students to focus on trying to get answers instead of trying to make sense of problems and situations. However, when learning is seen as an "it" that the students are supposed to "get," the student's role is seen more passively. Our language betrays this bias—the teacher "covers the material" and the students "absorb." Steven Leinwand, from the Connecticut State Department of Education, once told a joke at a mathematics teachers' convention about the Martian who came to visit American schools. In her report to the Martians, she said, "Teachers are people who work really hard and students are people who watch teachers working really hard!" In a similar vein, the NCTM has a button that reads: "Mathematics is not a spectator sport." Much of the joy of mathematics is examining a situation or problem and trying to understand it. My own experience with elementary school children and my own two children, Emily and Josh, has convinced me that young children naturally seek to make sense of the world that they live in and that for a variety of reasons many people slowly lose that curiosity over time.

Worthwhile Mathematical Tasks vs. Oversimplified Problems   Another common belief about learning mathematics is that we should make the initial problems as simple and straightforward as possible. I call this approach the Lysol approach—we clean up the problems to reduce the complexity (we use routine problems and one-step problems), we try to make the problems unambiguous, and we get rid of extraneous information (we make sure to use every number in the problems). However, one of the biggest drawbacks of this approach is that most of the problems that people solve outside of classrooms are complex and ambiguous, and part of the problem is to determine what information is relevant. A famous Sufi story nicely illustrates the difference between the two approaches. Nasruddin came home one night and found a friend outside on his hands and knees looking in the dirt. When Nasruddin asked, "What's happening?" his friend replied, "I dropped my keys." Nasruddin asked, "Where did you drop them?" His friend pointed: "Over there." Puzzled, Nasruddin asked, "Then why are we looking here?" The response: "Oh, the light is much clearer here." Although the light is much clearer when the problems are unambiguous, routine, and one-step, that is just not how students learn to think mathematically. One consequence of these richer problems (NCTM uses the term "worthwhile mathematical tasks") is that most students report that the problems are much more interesting than what they generally find in texts.

Owning vs. Renting   Another way of describing these two very different beliefs about learning mathematics that many of my students have found illuminating is to talk of the difference between owning and renting knowledge. Many students report that one year after finishing a mathematics course, they have forgotten most of what they learned—that is, if they retook the final exam, they would fail. This means that they rented the knowledge they learned: they kept it long enough for the tests, and then it was gone. However, this need not be so. If you are an active participant in the learning process and if the instructional strategies of the professor fit with how you learn, then you will own most of what you learn. When you take your methods course in a year or so, you will still remember the important ideas from this course. It is such a wonderful feeling to know that you got more than just 3 or 4 credits for the 100+ hours you spent over the course of the semester. If you plan to be an excellent elementary teacher (and I expect you all do), then you need to own what you learn.

An Integrated Approach   If we want students to retain the knowledge, then how they learn it has to be different from how it has been in traditional classrooms. One of my favorite words, and one you will find in the NCTM Standards, is *grapple*. I have found that if students grapple with problems and situations and try to make sense of them, they are more likely to retain what they learn in the process. I believe that a central part of the teacher's job is to select worthwhile tasks (Professional Standard 1) and to develop an environment that invites all students to learn and that honors differences in how they learn (Professional Standard 5). A classroom consistent with the NCTM vision does not look like a traditional classroom in which the teacher mostly lectures and demonstrates, and students generally take notes, ask questions, and answer questions the teacher asks. Rather, the classroom looks like an ongoing dialogue: the teacher presents a problem, possibly a brief lecture, and then the teacher facilitates the discussion around that question or situation (Professional Standards 2, 3, and 4), during which students naturally expect to make guesses (predictions and hypotheses) and try to explain their thinking and justify their hypotheses. Similarly, I believe that the textbook for this classroom must be different from a traditional book in which the important concepts and formulas are highlighted and in which the problems at the end of the chapter are generally pretty much like the examples.

In the introduction to NCTM's *Professional Standards for Teaching Mathematics,* the authors summarize major shifts in mathematics classrooms that have been called for by the NCTM. This passage nicely summarizes much of what I have just described.

We need to shift:

- toward classrooms as mathematical communities—away from classrooms as simply collections of individuals;

- toward logic and mathematical evidence as verification—away from the teacher as the sole authority for right answers;

- toward mathematical reasoning—away from merely memorizing procedures;

- toward conjecturing, inventing, and problem solving—away from an emphasis on mechanistic answer finding;

- toward connecting mathematics, its ideas, and its applications—away from treating mathematics as a body of isolated concepts and procedures.

If we can convince our students that fundamentally mathematics is a sense-making enterprise, then we change not only how much they work but also *how* they work. That is, if they see the relevance of the problems and concepts and develop confidence, they work harder; if they see the goal as sense making versus "getting it," they work differently. With this brief overview of what I mean by learning, let me explain how the two volumes work together.

Role of *Explorations* and Text

As just mentioned, if we believe that people learn better by grappling with richer problems (worthwhile mathematical tasks) as opposed to being shown how to do simpler problems, then this creates a whole host of changes in what the classroom looks like, what the instructional materials look like, and how students' knowledge is assessed. Basically, these explorations have been designed to have you grapple with important mathematical ideas. I have posed questions and tasks that call on you to discover fundamental mathematical concepts and structures: for example, percents can be seen as proportions. In the *Explorations,* you grapple with new ideas and concepts in a hands-on environment. The text then serves as a resource: a place where the formal definitions and structures are laid out (to be used generally after the *Explorations*), a place to refine your understanding, a place to self-assess, and of course, a source of homework problems. The explorations in this volume are generally open-ended and often have more than one valid answer, whereas the investigations (in the text) are generally more focused in scope and, though they generally have one correct answer, the discussions following the investigations show that they always lend themselves to different ways to arrive at that answer. I generally begin each new chapter with an exploration that, in learning theory terms, enables and encourages students to construct a scaffolding of major mathematical concepts. I then use the text as a resource where the students can see the concepts articulated in an organized way.

Looking back is a habit of mind that I want to encourage. A famous mathematics teacher proverb is "you're not finished when you have an answer; you're finished when you have an answer that makes sense to you." I model this notion explicitly in Chapter 1 in the *Explorations* and in the text, and you'll see it often in both volumes. The idea of reflection is new to many students, but getting into the habit of looking back is a great way to take ownership of this content. With this in mind, I hope you're looking forward to exploring elementary mathematics.

Foundations for Learning Mathematics

Y ou are about to begin what we hope is a very exciting course for you—a course in which you will come to a deeper understanding of the important mathematical concepts and ideas in elementary school mathematics. One of the major differences between this and other mathematics texts arises from a growing belief that students learn mathematics best by using explorations to investigate and understand mathematical concepts and procedures, rather than merely having the teacher explain the concepts and procedures. Both volumes have been designed to present you with what the National Council of Teachers of Mathematics (NCTM) has called "worthwhile mathematical tasks." Here are three important characteristics of worthwhile mathematical tasks.

First, a worthwhile mathematical task contains significant mathematics that will require the students to *think* (as opposed to memorize or simply follow a procedure). As a result of working on such a task, students will have a deeper understanding of concepts and develop mathematical tools described in Chapter 1 of the text.

Second, the task must be interesting to the students. Virtually all of the tasks in the two books not only are rich mathematically but also are very similar to problems that elementary schoolchildren do. Both the text and the website contain references where you can read articles or see children doing very similar problems!

Third, the task should be rich enough so that *all* students can be challenged by the task and *all* students can feel success in the doing of the task.

To the extent that a course is rich in worthwhile tasks, you feel your own mathematical ability growing; you feel the difference between "owning" the answers and "renting" the answers. When students own, they are more confident, which makes them more motivated, which makes them work both harder and smarter, which makes them own more, which increases confidence, which increases motivation, which. . .

As you do these explorations, please keep the NCTM Standards in mind and use the "4 Steps for Solving Problems" guidelines on the inside front cover of *Explorations* to help you when you get stuck. There is an old adage in teacher education that teachers teach not the way they were taught to teach but the way they learned. So *how* you learn will have a lot to do with how you teach!

Mathematics for Elementary School Teachers pp. 6, 21

EXPLORATION 1.1 — Patterns, Problem Solving, and Representations

These opening problems focus on several important tools that you will develop and use throughout the course.

1. *Patterns* Patterns is one of the "big ideas" of elementary school mathematics—recognizing patterns, describing patterns, and extending patterns. Young children practically bubble with excitement when they see patterns. Patterns are not just fun, they are also important mathematically. When you see patterns, you make connections, you see more deeply into the mathematical structure, and you become able to solve problems more skillfully.

2. *Representations* You will be asked to think about how you represent (that is, record and show) your work. In class discussions, you will see various representations that others have used, and you will see how thinking about representations can enable you to solve problems more skillfully.

3. *Making predictions and generalizing* An important way in which mathematics is used is making generalizations, which enable businesses and scientists to make predictions.

PROBLEM 1: Patterns in Pascal's triangle

We begin with a pattern that Blaise Pascal discovered in 1653. Mathematicians have found patterns from this triangle in many different areas of mathematics, from algebra to probability.

1. Look at the version of Pascal's triangle on page 19. What do you see? Write down at least four observations and patterns.

2. Select one pattern or observation, and describe it as if to someone on the phone who can see the triangle but who did not see this pattern or observation. Describe it so that the person can see it from your description.

3. Describe where the following sets of numbers can be found in Pascal's triangle. Some are obvious, some are hidden, and some are composed by adding cells together.

   a. Natural numbers: 1, 2, 3, 4, . . .
   b. Triangular numbers: 1, 3, 6, 10, . . .
   c. Square numbers: 1, 4, 9, 16, 25, . . .
   d. Fibonacci numbers: 1, 1, 2, 3, 5, 8, . . .
   e. Hexagonal numbers: 1, 6, 15, 28, . . .
   f. Powers of 2: 2, 4, 8, 16, . . .

PROBLEM 2: How many handshakes?

Determine how many students are in your class, including yourself. If each student shakes hands with every student, how many handshakes will there be?

1. Work on this problem alone for a few minutes. Refer to "4 Steps for Solving Problems" on the inside cover as needed.

2. Discuss your ideas. Describe new ideas that you like that arose from the discussion.

3. Now solve the problem.

4. After the class discussion, select a solution path that was different from yours. Explain that way of doing the problem, as if to a student who was not in class today.

5. On Tape 17 in the *Teaching Math: A Video Library, K–4* series, a fourth-grade teacher poses the following question to her class: If all 24 children in the class exchange Valentine's Day

cards, how many cards will be needed? Do you think the answer to this question will be the same as the number of handshakes there will be for 24 people? Why or why not?

6. If you see connections between this problem and parts of Problem 1, describe them.

PROBLEM 3: How many ways to make money and stamps

1. How many different ways can you make 25¢ with coins?

    Randomly thinking of different combinations will not help you to grow as a mathematics student. Thus, think about how you might be systematic (there are different ways) and think about how your representation of the problem might help you to find all of the combinations.

    a. Now find and show all the combinations in some kind of table.

    b. Describe at least three patterns you see in your table.

2. How many different ways can you make 50¢ with coins?

    Three key goals in this problem are to continue with your development in being systematic, to look at representations that will be more useful, *and* to think about how you can use the work from the 25¢ problem to answer this question.

    a. Find and show all the combinations in some kind of table.

    b. Describe at least three patterns you see in your table.

    c. Describe how you used your work in part (a) to solve this problem.

3. Here are several coin problems that have a different flavor. The goal is not simply to get the answer, but to think about tools that you can use other than just random guess–check–revise.

    a. How can you have 5 coins that add up to 55 cents? Is there another solution? Can you *prove* that there is or isn't?

    b. How can you have 9 coins that add up to 60 cents?

    c. How can you have 10 coins that add up to 60 cents?

    d. How can you have 21 coins that add up to one dollar? This was a Math Forum problem, for which student work can be seen at
    http://mathforum.org/library/drmath/view/59182.html

    e. Make up your own problem.

4. Stamps

    Think about how you can be systematic. In each problem, show all the combinations in some kind of table, and describe at least three patterns you see in your table.

    a. How many different ways can you make 37¢ postage with 10¢, 3¢, and 2¢ stamps?

    b. How many different ways can you make 40¢ postage with 8¢, 5¢, and 2¢ stamps?

5. Make up your own problem.

PROBLEM 4: How many rectangles?

1. **a.** This problem appeared in *Teaching Children Mathematics* by Robert Mann in the September 2004 issue, pp. 72–74. The Squares family saw a grid of solar panels that consisted of 25 small squares arranged in a 5 × 5 square. How many squares do you see in the panel?*

    b. Discuss patterns you see in this problem that would enable you to determine how many squares would be in a 6 × 6 panel. In an *n* × *n* panel.

2. **a.** Now let us extend this problem. Going back to the 5 × 5 square, how many different rectangles can you find in this figure?

    b. First, you need to determine whether you count the squares as rectangles and also what comprises "different." Write down what your class decided in language that makes sense to you.

    c. Go about determining the total number of rectangles carefully. Think about how you are representing your work so that you won't miss any rectangles.

    d. What patterns do you see in your work?

3. **a.** What if we had a 6 × 6 square? How many different rectangles can you find in this square?

    b. How can you use your work from the 5 × 5 square to answer this question?

    c. If you haven't already done so, make a table to record your findings.

    d. What patterns do you see in the table?

4. **a.** What if we began with a 7 × 7 square? How many different rectangles can you find in this square?

    b. Describe how you can use your work from the previous problem to answer this question.

    c. What patterns do you see in your table for this problem?

5. Predict how many rectangles are in an 8 × 8 square, just by looking at patterns in your tables. Explain your prediction.

EXPLORATION 1.2   Patterns in Multiplication

1.   Look for patterns in the problems below to enable you to predict the answers for (a) through (e). In each case, explain your prediction. If you need to make up more problems to see patterns, feel free to do so. Make up your own problem for (f), predict the answer, and justify your prediction.

23 × 99	2277
26 × 99	2574
34 × 99	3366
48 × 99	a.
148 × 99	14652
153 × 99	15147
175 × 99	b.
253 × 99	25047
264 × 99	c.
361 × 99	35739
384 × 99	d.
872 × 99	e.
? × 99	f.

2.   Look for patterns in the multiplication problems below to enable you to predict the answers for (a) through (f). In each case, explain your prediction. Make up your own problem for (g), predict the answer, and justify your prediction.

36 × 999	35964
47 × 999	46953
53 × 999	52947
68 × 999	a.
128 × 999	127872
163 × 999	162837
184 × 999	b.
245 × 999	244755
295 × 999	c.
1265 × 999	1263735
1358 × 999	1356642
1784 × 999	d.
2150 × 999	2147850
2185 × 999	e.
4850 × 999	f.
? × 999	g.

3.   Look for patterns in the multiplication problems below to enable you to predict the answers for (a) through (d). In each case, explain your prediction. Make up your own problem for (e), predict the answer, and justify your prediction.

25 × 101	2525
34 × 101	3434
56 × 101	a.
143 × 101	14443
157 × 101	15857
168 × 101	b.
243 × 101	24543
257 × 101	25957
286 × 101	c.
754 × 101	d.
? × 101	e.

4. Many interesting things happen when we multiply by 91.

    a. Fill in the blanks for (a)–(f) by applying patterns you have observed.

    b. Make up your own problem for (g).

33	91	3003
333	91	30303
3333	91	303303
33333	91	a.
44	91	b.
444	91	c.
4444	91	d.
2233	91	203203
2244	91	204204
2255	91	e.
3344	91	304304
4466	91	f.
?	91	g.

EXPLORATION 1.3   **Real-life Problems**

Each of these problems requires making assumptions and gathering some data in order to make a decision.

PROBLEM 1: How much will they save?

A couple bought a new coffee maker that makes individual cups so they can each have a cup of their own coffee in the morning. This coffee maker uses funnel filters instead of the circular-basket coffee filters used by larger coffee makers, so the couple bought a package of filters. However, they noticed that these filters were more expensive than the basket filters and that there was a big price difference between generic and brand-name filters: Brand-name funnel filters—40 for $1.50; generic basket filters—100 for 77¢. How much can they expect to save yearly if they switch from the brand-name funnel filters to the generic basket filters?

1.   First describe the assumptions you needed to make in order to solve the problem.
2.   Solve the problem and show your work.

PROBLEM 2: Which phone card?

A convenience store advertised two phone cards. Which card would you buy?
Card 1: 3.9¢ a minute with a 37¢ connection charge for each call.
Card 2: 7.9¢ a minute; no connection charge.

1.   First describe the assumptions you needed to make in order to solve the problem.
2.   Solve the problem and show your work.

PROBLEM 3: How much tape should I buy?

In order to revise and update each new edition of this book, I need to tape every page of the current textbook and explorations onto a sheet of blank paper, because these sheets are thinner than regular copy paper and get a bit frayed as the material goes from one person to another. I need to tape each side of the paper going up and down. How much tape will I have to buy?

1.   First describe the assumptions you had to make in order to solve the problem.
2.   Solve the problem and show your work.
3.   What if the publisher said I also needed to tape the top and bottom? Now, how much tape would I have to buy?

PROBLEM 4: How much soda should you buy for the party?

Let's say you are having a party, and you are expecting 120 people. Let us focus on the decision about how much soda to buy—you want to have enough soda for all the guests. The soda comes in 2-liter containers that cost 89¢. How much should you buy?

1.   First describe the assumptions you needed to make in order to solve the problem.
2.   Solve the problem and show your work.

EXPLORATION 1.4   Patterns and Proof

Making generalizations is an important part of doing mathematics. Looking for and analyzing patterns helps us to do this. At some point, we find we can make predictions. As we continue to explore, our predictions become more accurate and precise, and we also come to a deeper understanding of the structure of the problem, which enables us to get to the "why" that works its way to "proof." A major goal of these books is to reframe this notion of reasoning and proof so that justifying your work is something that you see value in doing.

PROBLEM 1: Darts

Variations of this problem are found in many elementary school textbooks. I used it with a group of fifth-graders with whom I worked once a week during the 1998–1999 school year.

1. If you have 4 darts and the dart board shown at the right, which of the scores below are possible scores, assuming that every dart hits the dart board?

     6   10   13   15   20   28

2. Predict what kinds of scores are possible and what kinds are not possible. For example, do you think a score of 18 is possible or not? Why?

3. Can you prove your generalization about what kind of numbers are possible and impossible?

4. What if you had only 3 darts? What scores would then be possible?

5. Make up and answer your own dart problem.

PROBLEM 2: Latin squares

Materials: Other Base Graph Paper (at end of book)

1. How many different ways can you insert four X's into a 4 × 4 array so that no row, no column, and no diagonal has more than one X? One solution is shown at the right.

2. Prove that the number of ways you found in 1. is true.

3. How many solutions can you find for a 5 × 5 array?

4. After the class has discussed the answer to part 3, describe patterns you see in the solutions for a 5 × 5 array.

PROBLEM 3: Diagonals of rectangles

Materials: At least 2 pages of Other Base Graph Paper (at end of book).

If we draw diagonals through grids, the number of squares that the diagonal will pass through can be predicted. See the examples below. The goal is to be able to say, "In an *L* by *W* rectangle, the diagonal will pass through *n* squares."

There are many ways that one can proceed to arrive at the generalization. Here is one path that can be helpful. It involves examining "families." For example, the 3-family begins with a 3 × 1 rectangle, and then the height is increased, systematically, one unit at a time. The table below shows the data for the first six members of the 3-family.

1.   Continue with the 3-family for several more cases, e.g., 3 × 7, 3 × 8, etc. Look for relationships between the length (*L*) and the width (*W*).

2.   Explore new families and look for patterns that can help you make generalizations. Predict the number (*N*) of squares that the diagonal will pass through, and explain how you arrived at your prediction. Check your predictions. If you are right, great; try another family. If a prediction is wrong, see if you can modify your prediction or understand why it didn't work.

3.   Summarize what you have learned from your work. If you have no generalizations, summarize your patterns and observations. If you have some generalizations, state them.

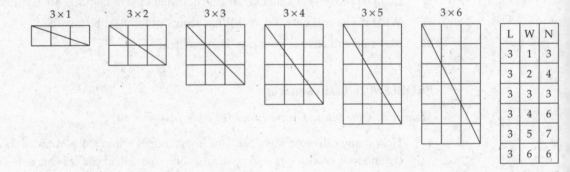

L	W	N
3	1	3
3	2	4
3	3	3
3	4	6
3	5	7
3	6	6

PROBLEM 4: Palindromes

Palindromes are numbers that read the same forward and backward. For example, 3663 is a palindrome. There is a process for creating palindromes from numbers that are not palindromes: Pick a number and reverse the digits. Add these two numbers together. If this number is a palindrome, we call the original number a 1-step palindrome. If not, take the new number, reverse its digits, and add these two numbers together. Continue this process until the sum is a palindrome. The example below shows why 68 is a 3-step palindrome.

$$
\begin{array}{r} 68 \\ + 86 \\ \hline 154 \end{array}
\qquad
\begin{array}{r} 154 \\ + 451 \\ \hline 605 \end{array}
\qquad
\begin{array}{r} 605 \\ + 506 \\ \hline 1111 \end{array}
$$

Explore all two-digit numbers from 11 to 99 and determine which numbers are 1-step palindromes, 2-step palindromes, 3-step palindromes, etc. As you go through the numbers from 11 to 99, look for patterns or observations that can shorten your work. For example, I know this is a 1-step palindrome because . . . , I know this is a 2-step palindrome because. . . .

1. Make a table like the one below that lists the already palindromes, 1-step palindromes, 2-step, etc.

Already	1-step	2-step	3-step	4-step	5-step
11	14	48			

2. Describe the characteristics of members of each family. In some cases, all of the members of that family have the same characteristics. In some cases, you will have to say—these x-step palindromes have this characteristic and these y-step palindromes have this characteristic.

3. In the table below, color the already palindromes with one color; color the 1-step palindromes with another color; color the 2-step palindromes with another color, etc. What do you see?

11	12	13	14	15	16	17	18	19	20
21	22	23	24	25	26	27	28	29	30
31	32	33	34	35	36	37	38	39	40
41	42	43	44	45	46	47	48	49	50
51	52	53	54	55	56	57	58	59	60
61	62	63	64	65	66	67	68	69	70
71	72	73	74	75	76	77	78	79	80
81	82	83	84	85	86	87	88	89	90
91	92	93	94	95	96	97	98	99	100

*Mathematics
for Elementary
School Teachers*
p. 22

EXPLORATION 1.5   Magic Squares

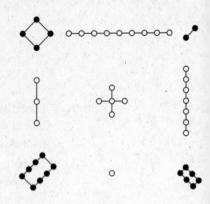

Magic squares have fascinated human beings for thousands of years. The oldest recorded magic square, the Lo Shu magic square, dates to 2200 B.C. and is supposed to have been marked on the back of a divine tortoise that appeared before Emperor Yu when he was standing on the bank of the Yellow River. In the Middle Ages, many people believed magic squares would protect them against illness! Even in the twenty-first century, people in some countries still use magic squares as amulets.

As a teacher, you will find that many of your students love working with magic squares and other magic figures.

PART 1: Describing magic squares and finding patterns

1. The definition of a magic square is that the sum of any row, any column, and each diagonal is the same, in this case 15. Look at the square and write down anything that you observe in this square—relationships between numbers in rows or columns or diagonals, patterns in how the numbers are arranged, even/odd, etc.

8	1	6
3	5	7
4	9	2

2. Compare your observations with others in your group.

PART 2: Patterns in all 3 × 3 magic squares

1. You will find eight different 3 × 3 magic squares on page 15. Look at these magic squares carefully and then write down

    a. Patterns that are found in all of the magic squares

    b. Patterns that are found in some of the magic squares

2. Compare your observations with others in your group.

3. Now that you have seen nine different magic squares, describe how you could make a completely new magic square.

PART 3: Using algebra to describe magic squares

It is possible to use algebraic notion to describe the structure of all 3 × 3 magic squares just like we can compute the area of all rectangles by multiplying length by width.

1. What if we called the middle number of any magic square m? Take the first two magic squares on page 15, and represent the other numbers in each square in terms of the middle number. For example, in the first square, the middle number would be m, and the number to its right would be $m + 3$. What patterns do you notice? What similarities do you notice between the two magic squares?

2. One way of making a general representation of 3 × 3 magic squares is to use two more variables. Using the patterns and observations from (1) above, now represent the first two magic squares on page 15 using m, x, and y.

3. At this point, you have a solution generated either from your group or from the whole-class discussion. Think about your work and discussions and then read the following quotation: "Mathematics is often considered a difficult and mysterious science, because of the numerous symbols which it employs. . . . [T]he technical terms of any profession or trade

are incomprehensible to those who have never been trained to use them. But this is not because they are diffi-cult in themselves. On the contrary they have invariably been introduced to make things easy. So in mathemat-ics, granted that we are giving any serious attention to mathematical ideas, the symbolism is invariably an immense simplification."[1] Does it help you to see the use of symbols in a new light?

PART 4: Further explorations

1.   Determine the missing numbers in the magic squares below.

7		
5	9	13

17	13	
15		19

2.   How many different 3 × 3 magic squares can you make starting with these two numbers?

3		
		7

3.   Below are questions about four possible transformations of a 3 × 3 magic square. In each case, write your initial guess and your justification before you test your guess. Then test your guess. If you were correct, refine your jus-tification if needed. If you were wrong, look for a flaw or incompleteness in your reasoning.

   a. If you doubled each number in a magic square, would it still be a magic square?

   b. If you added the same number to each number, would it still be a magic square?

   c. If you multiplied each number in a magic square by 3 and then subtracted 2 from that number, would it still be a magic square?

   d. If you squared each number in a magic square, would it still be a magic square?

4.   Look back on your work from different 3 × 3 magic squares, and answer the following questions:

   a. Is there a relationship between the magic sum and whether the number in the center is even or odd?

   b. Divide the set of magic squares into two subsets: those in which the nine numbers are consecutive numbers (such as 10–18) and those in which the nine numbers are not consecutive numbers. Are there any other differ-ences between these two sets of magic squares?

3 × 3 Magic Squares and Templates for EXPLORATION 1.5

18	33	15
19	22	25
29	11	26

7	17	3
5	9	13
15	1	11

6	10	11
14	9	4
7	8	12

7	26	6
12	13	14
20	0	19

14	28	12
16	18	20
24	8	22

9	26	7
12	14	16
21	2	19

10	21	8
11	13	15
18	5	16

6	13	8
11	9	7
10	5	12

EXPLORATION 1.6   Magic Triangle Puzzles

Many children enjoy problems like the ones in this exploration, and the problems contain great opportunities to develop mathematically. In addition to providing computation practice, they also help children develop more powerful problem-solving strategies and develop their reasoning abilities.

PART 1: Magic triangles

The task is to determine the numbers that belong in the circles so that the sum of the numbers in any two circles equals the number between them. The first problem is worked out for you below.

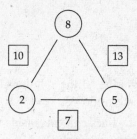

1.   Now solve the puzzles below. As you play with these problems, look for patterns that might unlock the problem so you don't have to do random guess-check-revise every time:

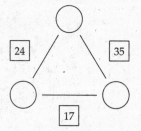
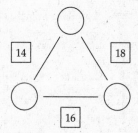
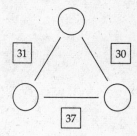

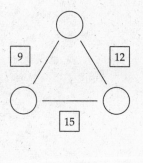
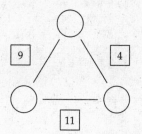
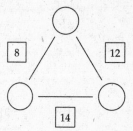

2.   Once you have a method that will solve any triangle quickly, write your directions for your method and give them to someone not in this class. Ask them to solve the triangle problems just on your directions. If they get stuck, ask them what part of your directions is confusing. Straighten out that confusion and revise your directions.

3.   Give the revised directions to another person. How did they do?

4.   What did you learn about writing clear directions?

5.   Make up some of your own probelms.

PART 2: A different kind of magic triangle puzzle.

This exploration is adapted from "Building a Community of Mathematicians" in the April 2003 issue of *Teaching Children Mathematics.**

1. Using the numbers 1 through 6, place numbers in each circle so that when you add the three numbers that make up each side of the triangle, you get the same sum in all three cases. How many different solutions can you find?

2. Look again at all the solutions that your class found. What do you see that is interesting? What patterns do you see?

3. Now draw line segments inside the puzzle so that the numbers in the middle of the sides of the triangles are connected. You can now break each puzzle into three small triangles, each comprising three numbers. What patterns or relationships do you see in these triangles?

Pascal's Triangle

CHAPTER 2

Fundamental Concepts

The concepts of sets, functions, and numeration permeate elementary school mathematics. The goals of the explorations in this chapter are for you to recognize and be comfortable with these fundamental concepts. Throughout the book, we will run into situations in which ideas related to sets and functions enable us to communicate more easily and help us to understand and solve problems. Similarly, a deeper understanding of numeration will enable you to develop what the NCTM calls mathematical power.

SECTION 2.1   Exploring Sets

As you will discover in your methods course, the concept of sets (and subsets) is an important aspect of the development of logical thinking in young children.

Do you know that Jean Piaget, a Swiss psychologist, used set concepts to help determine the extent to which a young child could reason abstractly? He learned how children's thinking developed by conducting "interviews" with children, one of which is illustrated below. A child is given a set of objects that are all mixed together. The child is then asked to separate the objects into two or three groups so that "each group is alike in some way." Do this yourself and record your different solutions.

The following explorations are designed to enhance your ability to work with sets and subsets so that when you teach, you will be able to recognize and nurture this important aspect of children's "mathematical" thinking.

*Mathematics
for Elementary
School Teachers*
p. 60

EXPLORATION 2.1   Understanding Venn Diagrams

We will explore Venn diagrams as a tool for gathering and reporting data about different subsets of a set.

Consider the following Venn diagram, in which the areas listed below represent various sets of students:

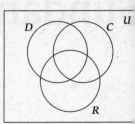

- *U* represents the set of students in this class.
- *D* represents the set of students who have at least one dog.
- *C* represents the set of students who have at least one cat.
- *R* represents the set of students who have at least one rodent, such as a rat, mouse, gerbil, hamster, or guinea pig.

1.   Write your initials in the region in which your household belongs. If you live in a residence hall, think of your parents' home or your previous home. Then discuss the diagram with other members of your group. Each member should show where his or her name was placed and why.

In Steps 2 through 6, refer to the class Venn diagram on an overhead projector or blackboard.

2.   Which subset has more members: students who have cats and dogs, students who have cats and rodents, or students who have dogs and rodents? How did you figure this out?

3.   What fraction (percentage) of the class has at least one dog?

4.   How would you describe in words the subset of the class whose names are outside all three circles?

5.   Describe a different subset of the class:

   a. In words

   b. With a Venn diagram

6.   How would you describe the subset of the class that has no pets?

EXPLORATION 2.2   Gathering and Interpreting Data

This exploration involves collecting data and then using set ideas to analyze those data.

Many writers have described our era as the Computer Age. Many items that are found in most homes today either did not exist 20 years ago or were found in only a small percentage of homes. Think of the household you belong to; if you live in a residence hall, think of your parents' home or your previous home. Does that household have a computer, a DVD player, or an mp3 player?

1. *Ask the questions*   If we consider the students in this class as a subset of the set of college students, what questions might you ask concerning ownership of these items?

    a. Write the questions that you would like to ask, and then discuss the questions.

    b. After the class discussion, write the questions that will actually be asked.

2. *Gather the data*   Determine how you will gather the data in order to answer the questions you are asking.

    a. Write your proposed method.

    b. After the class discussion, write the method that will be used. If this method is different from the one in part (a), discuss and then describe the advantages of the method that the whole class chose over the method that you described in part (a).

3. *Answer the questions*

    a. In your group, discuss how you can represent the information that you gathered in order to answer the questions.

    b. Prepare a brief report that includes your answers to the questions, a description of how you determined the answers, and your representation of the information (a chart or diagram).

SECTION 2.2   Exploring Algebraic Thinking

The concept of functions is one of the "big ideas" of mathematics. Although functions are not explored formally until high school, the foundation is laid in elementary school by exploring patterns. We see patterns almost everywhere, both in everyday life and in work situations, and children love playing with patterns. We connect this natural fascination with patterns to mathematical thinking by focusing on recognizing, describing, and extending patterns.

The notion of analyzing patterns and functional relationships has many real-life applications. One of the most poignant has to do with finding missing children. In one case, researchers helped investigators find two children who had been abducted by their father and had not been seen in eight years. Scott Barrows and Lewis Sadler developed a computer program that enabled them to take photographs of the five- and seven-year-old girls and create pictures of what they should look like eight years later. "The pictures showed up on TV one night and within 10 minutes a national hot line was getting calls from neighbors . . . by 7:30 the next morning the girls were in police custody."[1]

How had the researchers been able to do this? It seems that facial bones change in predictable ways throughout childhood. The two researchers determined relationships for 39 facial dimensions. By recognizing and analyzing the most important patterns, they were able to predict quite accurately what the girls would look like eight years later. In the explorations and investigations in this chapter, you will explore patterns and see how the concept of functions is related to understanding patterns.

Mathematics for Elementary School Teachers
p. 70

EXPLORATION 2.3   Exploring Equivalence

Young children tend to think of the equal sign as "the answer is," for example, $3 + 5 = 8$. If you presented this problem to a first or second grader, $8 + 2 = \square + 4$, they would be likely to say that the problem doesn't make sense or simply "you can't do that." In this exploration, we will use the image of balance to emphasize the equal sign as the equivalence of the quantities on both sides. This enlarged vision of equals is an essential foundation for developing algebraic thinking.

1. Let us begin with problems involving qualitative rather than quantitative thinking. Qualitative thinking requires you to think more verbally and to look at relationships as opposed to just knowing how to get the answer. The problems are adapted from *Balancing Act: The Truth Behind the Equals Sign,* by Rebecca Mann, in the September 2004 issue of *Teaching Children Mathematics.**

   Think of the following shapes as representing different weights.

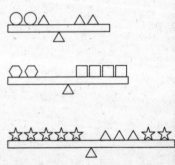

   a. In the picture at the right, two circles and a triangle balance two triangles. In this case, what can you say about the weights of the circles and the triangles?

   b. In the picture at the right, two hexagons balance four squares. What can you say about the weights of the hexagons and squares?

   c. In the picture at the right, five stars balance three triangles and two stars. What can you say about the weights of the stars and triangles?

---

*Reprinted with permission from *Teaching Children Mathematics*, copyright © 2004 by the National Council of Teachers of Mathematics.

2. Solve both of the problems below by a means other than writing two equations and solving them. In this way, you will be more likely to solve the problems as young children would.

   a. Two students are at a store having a clearance on CDs and DVDs.

   All the sale CDs are the same price and all the DVDs are the same price.

   How much does a CD cost? How much does a DVD cost?

   One student bought 2 CDs and 1 DVD for $25.

   The other student bought 1 CD and 2 DVDs for $23.

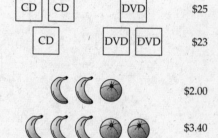

   b. Tasha bought 2 bananas and 1 orange for $2.00.

   Noah bought 3 bananas and 2 oranges for $3.40.

   How much does each fruit cost?

   $2.00

   $3.40

3. Children's solution paths to the first problem below was discussed in *Responses to the "Guess the Weight" Problem* on pp. 31–33 in the September 1999 issue of *Teaching Children Mathematics*. As before, solve the problems by a means other than writing and solving equations.*

   a. In the problem below, each of the shapes has a certain weight. Determine the weight of each shape.

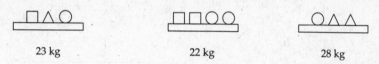

   23 kg   22 kg   28 kg

   b. In the problem below, each of the shapes has a certain weight. Determine the weight of each shape.

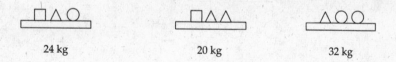

   24 kg   20 kg   32 kg

---

*Reprinted with permission from *Teaching Children Mathematics*, copyright © 1999 by the National Council of Teachers of Mathematics.

4.   You have seen mobiles hanging from ceilings. Imagine mathematical mobiles where eac
side balances. For example, in the mobile at the left, if the right circle has a weight of 4, the
the left circle would have to have a weight of 4 for the mobile to balance. (For our purpose
here, we will ignore the weight of the rods that connect the weights.) In the mobile at th
right, the weights of the two circles at the left would have to add to 4. That is, one could hav
a value of 3 and the other a value of 1, or they could each have a value of 2.

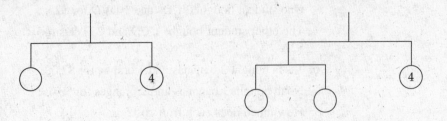

Determine the value of each weight to make the following mobiles balance.

a.

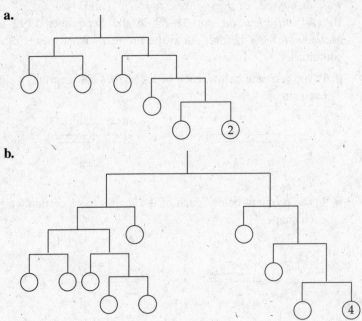

b.

c. Make up a problem yourself.

For similar problems, see "Solutions to Using Your (Number) Sense of Balance" i
Teaching Children Mathematics, March 2009, pp. 390–393.

EXPLORATION 2.4 Relationships Between Variables

This exploration is adapted from an article by Leah McCoy called "Algebra: Real-Life Investigations in a Lab Setting."[2] Although McCoy used these investigations with middle-school students, many of them can be adapted for use in most elementary grades.

The key idea here is to collect data and then look for relationships between the two variables so that you can see a trend or pattern that enables you to predict the future! This is one of the primary reasons why functional relationships are so important. When the relationship between two variables is a functional one, we can predict the future. A simple function is the hourly wage. For example, if you are being paid at the rate of $8.00 per hour, you know that your earnings will be determined by the function Earnings = 8 times the number of hours you work.

Your instructor will assign your group one of the questions below. Each group will perform the assigned exploration until you feel you have enough data to answer the question, and then each group will report its findings to the whole class.

The Explorations

1. What is the relationship between the number of people and the time it takes to make a wave? Start with a group of five students and make a wave (as is done at sporting events). Collect more data with larger groups of people.

2. What is the relationship between the number of weights and the length of a rubber band? You will be given a paper clip, a rubber band, and a number of weights, such as 10 heavy washers from a hardware store. Open the paper clip so that it will be able to hold the weights, and slip the paper clip on the rubber band. Place the weights, one at a time, on the paper clip and measure the length of the rubber band.

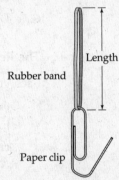

Rubber band

Length

Paper clip

3. What is the relationship between the number of dominoes in a run and the time it takes for them all to fall? Make a domino run and determine how long it takes for all the dominoes to fall. Repeat the experiment with different numbers of dominoes. *Note:* You will keep the distance between dominoes constant.

4. What is the relationship between the number of crackers eaten and the time before one can whistle? One person will be given x small crackers (such as oyster crackers or goldfish). How long after eating the crackers is it until the person can whistle? Repeat the experiment with different numbers of crackers.

5. What is the relationship between the height of an object and the length of its shadow? Go outside and measure the heights of various objects (including people) and the lengths of their shadows.

6. What is the relationship between the length of a pendulum and its period? Make a simple pendulum with string or fishing wire and a weight, such as a nut or washer from a hardware store. The length of the pendulum is the distance from the top of the string to the center of the weight. The period of the pendulum is the time it takes to make one full sweep, back and forth. Determine the period by swinging the pendulum 10 times and then dividing by 10.

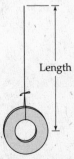

Length

7. What is the relationship between the number of drops of ink on a paper towel and the diameter of the circle formed by the ink? You will be given some paper towels, a bottle of ink, and a dropper. Carefully drop one or more drops of ink onto the paper towel and then measure the diameter of the circular region that is formed. Repeat the experiment with different numbers of drops.

The Procedures

Each group should go through the following steps in order to carry out its exploration.

1. Make sure everyone understands the problem.

2. Discuss how the data will be measured and recorded—in inches, in centimeters, with ruler or stop watches, and so on.

3. If precision is an issue, do some practice runs to make sure you can record the data as precisely as needed.

4. Collect the data.

5. After you have collected enough data so that you feel you see the relationship, select the next point where you would collect data and predict the outcome. Explain your prediction. Then collect the data. If your prediction was accurate, great. If not, discuss the difference between your prediction and the actual data, and repeat this step.

6. Determine your answers to the question and then make your report.

The Reports

Each group's report should include the following:

a. A statement of your problem/question.

b. A statement of what you did (your design).

c. A chart or overhead containing the following information or an explanation of why some of the information is not relevant to your situation:

- The independent variable (and unit)
- The dependent variable (and unit)
- A table of your data
- A graph
- An equation expressing the relationship between the variables
- The relationship between the variables, expressed in words

d. Your primary conclusions, including your degree of precision and your level of confidence in your conclusions. Describe any factors that limit your degree of precision and/or limit your ability to generalize.

e. A description of your biggest obstacle and how you overcame it.

*Mathematics
for Elementary
School Teachers
pp. 76, 78*

EXPLORATION 2.5   Connecting Graphs and Words

PART 1: Walking across campus

Because change is so concrete, the context of change is almost universally used as a beginning place for developing algebraic reasoning in elementary schools. In this exploration, we will start with a quantitative examination of change and then move to a qualitative examination. This work will enable you to grapple with important algebraic ideas.

Below are data collected as Keisha moves across campus.

Time (seconds)	0	1	2	3	4	5	6	7	8	9	10
Distance (feet)	0	4	8	12	16	16	16	20	26	35	48

1. Make a time–distance graph for this situation: The *x* axis is time, and the *y* axis is the distance from the starting point.

2. **a.** How does the slope of the graph correspond to Keisha's speed?

   b. What speed does a horizontal line represent?

3. Write a brief description of Keisha's walks shown in the four graphs below. Two versions of graphs (b), (c), and (d) are given. The top version represents changing speed instantaneously, like some professional athletes seemingly do. The bottom version represents the more realistic changing from one rate to another.

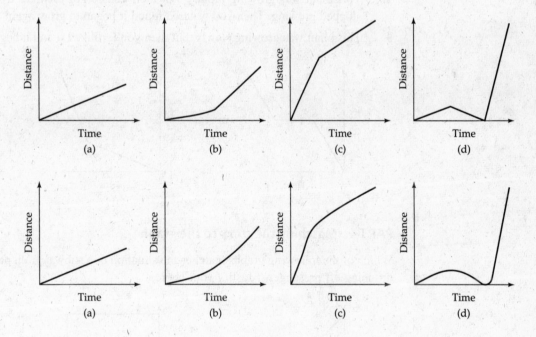

PART 2: Driving to work

The graph below is a representation of Felipe's drive from home to work. Write a short story that describes his trip.

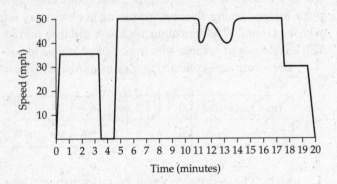

PART 3: Growing plants

Below are three scenarios and three graphs. Match each scenario to the correct graph and explain why that graph is correct.

1. This plant was growing rapidly in the sun, and then you moved it to a shady spot and it began to grow more slowly.

2. This plant was growing rapidly, but then you forgot to water it for a few days and so it stopped growing. Then you watered it and it began to grow again.

3. This plant was growing slowly, but then you fertilized it and it began to grow more rapidly.

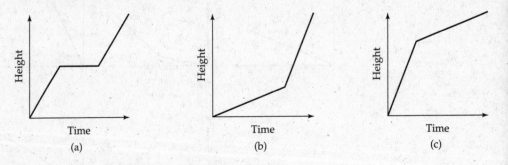

PART 4: Matching the story to the graph

Which of the following graphs matches the situation of a subway train pulling into the station and dropping off passengers? Justify your choice.

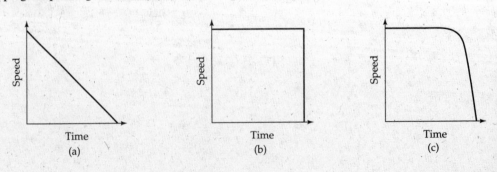

PART 5: Eating popcorn

Below are graphs representing four different ways in which people eat popcorn. Describe each person. Make a sketch that describes how you eat popcorn and give a brief verbal description that matches the graph.

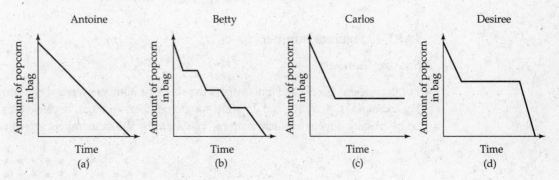

PART 6: Make a graph to match each story

Below are four situations. Sketch a graph that represents each story and briefly explain your graph.

1.   The amount of daylight (not sunlight) in your town over the course of a year.

2.   A graph of a typical person's height over the course of his or her lifetime.

3.   The amount of coffee needed for a pot of coffee, from 1 cup to many cups.

4.   The temperature of water in a pot from the time it is put on the stove until it is boiling.

5.   Below are three situations. Sketch a graph that represents each story.

   a. You are speeding on a highway, you get stopped by a police officer and fortunately get a warning, and then you resume your trip but at a slower rate.

   b. Represent the tortoise and the hare fable graphically. The independent variable is time, and the dependent variable is distance traveled.

   c. You planted a seed in a pot. Once it sprouted it grew quickly for a while and then more slowly.

*Mathematics
for Elementary
School Teachers*
pp. 79, 80

EXPLORATION 2.6   Growth Patterns

The problems in this exploration are "growing patterns" because they represent situations in which something is growing at a predictable rate. Growing patterns permeate everyday life and thus have found their way into most elementary schools.

PART 1: Figurate numbers

Square Numbers

For the ancient Greeks the amount and the shape of numbers were very connected in their minds. For example, 1, 4, 9, 16, and 25 were square numbers to them, because they could be represented by placing dots in a square pattern. The first few square numbers are shown below.

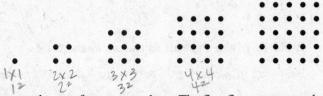

1.   Let us look now at the set of square numbers. The first five square numbers are shown above.

   a. How many dots does the 6th square number contain?

   b. How many dots does the 10th square number contain?

   c. How many dots does the nth square number contain?

Triangular Numbers

Look at the set of triangular numbers shown below. This problem is more difficult than the previous one because simply seeing the pattern does not easily generate a formula that enables you to predict the number of dots in the nth triangular number.

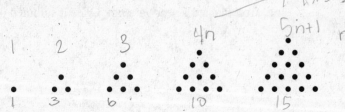

2.   How many dots are in the nth triangular number? Explain your prediction and how you came to your conclusion.

   If you have trouble with this question, look back in Chapter 1 to find any problems that connect to this question. Or represent the numbers on graph paper with squares instead of dots (see the figure at the right) and connect the question to what part of a complete rectangle has been made.

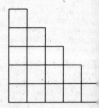

Relationships Between Figurate Numbers

3.   *Interpreting a new notion*   The following formula expresses *one* way in which square and triangular numbers are related: $S_n = T_{n-1} + T_n$

   a. Translate this equation into English.

   b. Exchange your translation with a partner and discuss each of your translations until you both feel satisfied that both phrasings work.

   c. Justify this relationship—that is, explain why it is true. Write your first draft of a justification.

4. ***Translating words into notation***

   a. Translate this sentence into notation: The square of any odd number is one more than eight times a specific triangular number.

   b. Exchange your notation with a partner and discuss each person's work.

Other Figurate Numbers

5. Look at the pentagonal numbers below.

   Describe patterns that you observe and how you can predict the 5th, 6th, and nth pentagonal numbers.

6. Look at the hexagonal numbers below.

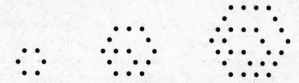

   Describe patterns that you observe and how you can predict the 5th, 6th, and nth hexagonal numbers.

7. Make a table like the one shown below. Write in your expressions for the value of the nth triangular number and the nth square number. Determine the value of the nth pentagonal number and the nth hexagonal number. Show your work.

Type of number	Value of the nth term
Triangular	
Square	
Pentagonal	
Hexagonal	
nth figurate number	

8. Can you predict the value of the nth number for *any* figurate-number family, such as the nth octagonal number?

PART 2: Squares around the border

The following problem is a popular growing pattern problem in elementary schools because it is so rich. The February 1997 issue of *Teaching Children Mathematics* has a whole article describing how this problem can be adapted for grades K–2, for grades 3–4, and for grades 5–6. I encourage you to look it up in your library. I have also seen variations of this problem at several conferences and in other articles. With young children, the problem is often presented with blue and white tiles (that you can get at a tile store). The blue tiles (in the center) represent the water in the pool, and the white tiles represent the border around the pool.

1.   Below you can see the first five figures in this growth pattern. You will count the number of squares in the border of each figure, and then, from the patterns you see and the observations you make, determine the number of squares around the border of the *n*th figure. As in Part 1, we will break the problem into a series of smaller steps.

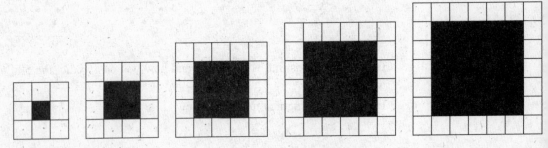

a. How many squares are there in the border of the first figure? of the second figure? of the third figure? of the fourth figure? of the fifth figure? What patterns do you see, in the figure and/or from how you counted the number of squares? Draw additional figures if you feel that would help.

b. Your instructor will now have you share your observations with your partners or the whole class.

c. Working at first on your own and then with your partners, try to make use of observations and patterns to determine the number of squares in the border of the *n*th figure.

2.   One of the themes of the book is the notion of multiple representations. Often, we see new aspects of a situation when we examine it from another perspective. We will do that here— the other perspective is to see what graphs tell us about the relationships among the figures, the number of squares around the border, and the number of squares in the middle.

a. Make a table like the one on the next page. Fill it out and then make two graphs. In the first graph, the independent variable will be the number of the figure (1st, 2nd, 3rd, etc.), and the dependent variable will be the number of squares around the border of that figure. In the second graph, the independent variable will again be the number of the figure, but the dependent variable will be the number of squares in the middle of that figure.

b. Record any observations you make from the two graphs.

c. How would you describe, in words, the growth in the number of squares around the border and the growth in the number of squares in the middle?

Figure number	Number of squares around the border	Number of squares in the middle	Notes on how you determined this
1			
2			
3			
4			
5			
6			
7			

Extensions

As you are finding, many rich problems have more to offer even after you have found the answer. They are extendible! Let us examine three extensions of the squares-around-the-border problem.

3.   **a.** Determine the fraction of the total area that is contained by the "pool" (inner shaded region) for the first five figures on page 34.

   b. Describe the changes you see in the fraction.

   c. What fraction will be pool in the nth figure?

   d. Make a graph of this situation: The independent variable is the number of the figure, and the dependent variable is the fraction of the area that is pool. Describe this function in words.

4.   The growth pattern shown below adds one more layer to this problem. We have the white outer border and a shaded inner border. How many squares would make up the nth outer border (white squares)?

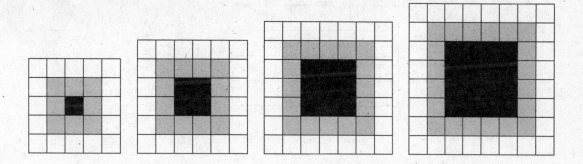

5.   This extension comes from realizing that we don't have to have a square "pool" and a square "border." How many squares are in the border of the nth figure for the growth pattern shown below?

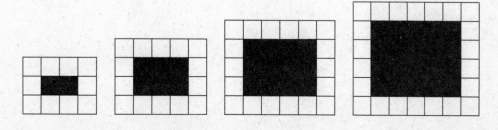

PART 3: Growth patterns with pattern blocks

Pattern blocks can be found in most elementary classrooms and are a versatile and enjoyable manipulative. Children will build all sorts of interesting patterns with pattern blocks. Let us examine two patterns.

1.   Begin with one yellow hexagon. Each side of the hexagon is 1 inch, and thus the hexagon has a perimeter of 6 inches. Now place another hexagon next to the first. The perimeter of this shape is 10 inches. Continue this pattern. The question: What will be the perimeter of n hexagons in a row?

2.   Now imagine building a patio with pattern blocks, beginning with the yellow hexagon. Next we will build a ring around the yellow hexagon, making a larger hexagon. Next we will build another ring, making an even larger hexagon. Let's say the cost of the first hexagon is $6. If we make one ring, the ring will cost $18 and the total cost for the whole figure will be $24. If we continue to build, how much will the nth ring cost, and how much will the total structure cost after we build n rings?

Ring	Ring cost, $	Total cost, $
0	6	6
1	18	24
2		
3		
...		

Initial block   First ring

SECTION 2.3 Exploring Numeration

One day an elementary teacher, Georges Ifrah, was asked by one of his students, "How did numbers start? When did people learn to count?" At the time, all he could say was "I don't know" and "a long time ago." To answer those questions, he did quite a bit of research and wrote a fascinating book called *From One to Zero: A Universal History of Numbers*. Historical records show that many different kinds of number systems were created as humans' understanding of mathematics increased. You will spend some time creating your own numeration system so that you will be able to understand more deeply how we use numbers.

EXPLORATION 2.7 Alphabitia

Imagine that you are a member of a small tribe that lived thousands of years ago, when people were making the transition from being hunter-gatherers to becoming farmers. You have a numeration system that is alphabetically based, so you are called Alphabitians. As is true of many other ancient peoples, your numeration system is finite. For any amount greater than Z, you have no symbol; you just call that amount "many."

Amount .... A B C D E ...Z many
Alphabitian numeral

PART 1: Inventing a new system

Now that your tribe has settled down, you have "many" sheep and "many" ears of corn. Without an adequate numeration system, figuring out how many more sheep you have this year than last year and determining each family's share of the corn harvest are very tedious. Recently a young woman in your tribe had excitedly announced that she had invented a new counting system with which she can represent any amount using only the symbols A, B, C, D, and a new symbol she calls zero and writes as 0. Unfortunately for your tribe, this young woman died on a hunting trip. However, she left behind some artifacts that she was going to use to help you learn the new system. These artifacts are called flats, longs, and units.

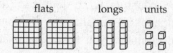

flats   longs   units

Because the visionary member of your tribe is no longer with you, it is up to you to invent the new numeration system that your tribe desperately needs. That is, you need to develop a system that lets you represent *any* amount using only the symbols A, B, C, D, and 0.

1. Cut out a number of flats, longs, and units from graph paper or use the manipulatives provided by your instructor.

2. Take some time to sit down with your partners and create a numeration system using only the symbols A, B, C, D, and 0.

3. Some groups will find that one member proposes a system that makes sense to everyone quickly. Other groups will discuss and debate two or more systems before finally deciding on one system. After you have explored different alternatives, answer the following questions before proceeding to Part 2.

a. Does your system make sense to every member of the group?

b. Imagine that you will be explaining your system to the council of the elders. How would you explain your system to them?

c. If your group decided between two (or more) possibilities, what made you choose one system over the other(s)?

PART 2: Communicating different systems

At a time specified by your instructor, each group will post its system on the wall, and each student will have time to look at the other systems.

1. Make a group poster. Each group's poster needs to contain the following:

   • The name of your group (feel free to be creative and playful).

   • The table on page 41, showing your symbols for at least the first Z numbers.

   • Directions that will help other students to understand the system (*Remember:* You will not be there to explain it to other students.)

   • A frank and honest assessment of the advantages and the disadvantages or limitations of your new system.

Old	New	Picture
A		.
B		:
C		:
D		:
E		I
F		I.
...		
X		
Y		
Z		

2. *First discussion of the systems: observations and questions*
After students have had a chance to examine the different systems, there will be a class discussion in which you will be asked to share your observations and questions about other systems.

   a. Jot down observations or questions that you would like to remember.

   b. What did you learn from this discussion?

3. *Taking different systems for a test drive* After examining all the systems, select the two systems that you like best and take them for a "test drive." You may choose the system your group developed as one of the two systems. Each test drive must include the following tasks:

   a. Try counting up to double Z (that is, the amount equivalent to Z + Z in the original Alphabitian system). Record your symbols and describe any difficulties you had and any uncertainties you have. For example, you may see two reasonable possibilities for a certain amount. Reminder: The goal here is "sense making" as opposed to "just do it." The more you get into this role play, the better you will understand the difficulties young children have when learning to count in our numeration system!

   b. Make up and solve some simple story problems involving addition (e.g., new sheep born), subtraction (e.g., selling sheep), multiplication (e.g., planting rows of trees), and division (e.g., dividing ears of corn among several families). Please stay in your role as Alphabitians. Among other things, this means forgetting the computation procedures you know as Americans. Thus, when you divide, think how you (as an Alphabitian) might determine the answer to the question in your story problem.

   c. Evaluate each system. First summarize its strengths and its weaknesses or limitations. Then make any recommendations for improving the system.

4. *Second discussion of systems: choosing one system* At this time, your class will decide on one system that you will all learn and use. You might select one of the systems created, or you might combine two systems (each with certain advantages) and then select the combined system.

   a. As different students nominate a system and then debate the advantages and limitations of that system, make notes about any points that you want to remember.

   b. What did you learn from this discussion?

PART 3: Learning about the new system

1. **a.** After the class selects the system that it will use, take time to learn that system. You may want to review the "4 Steps for Solving Problems" before you begin.

   b. Once your group feels that each member understands the new system, take a few moments to jot down structures of this new system that help you to count quickly and confidently.

   For Steps 2 and 3, take some time to answer the following questions. Answer them using manipulatives, diagrams, or symbols.

2. What number comes after each of the following? Briefly explain how you answered each question.

   a. AA   **b.** BD

   c. BAD   **d.** ABD

3. What number comes before each of the following? Briefly explain how you answered each question.

   a. D0   **b.** B00   **c.** D0D0

4. Compare your responses to Steps 2 and 3 with those of other members of your group.

   Take some time to analyze mistakes. Were they just careless, or do they point to some aspects of this system about which you are still fuzzy?

   Your instructor may have you do one or more additional worksheets.

5. Describe the most important learnings that you derived from this exploration. In each case, first describe what it was that you learned, and then describe how you learned it.

6. Describe something in this exploration that you are still not clear about.

Alphabitia Table for EXPLORATION 2.7

Old symbol	New symbol(s)	Picture of what the amount looks like	Explanation
A		.	
B		:	
C		⋮	
D		⋮	
E		|	
F		|˙	
G		|:	
H		|:	
I		|:	
J		||	
K		||˙	
L		||:	
M		||:	
N		||:	
O		|||	
P		|||˙	
Q		|||:	
R		|||:	
S		|||:	
T		||||	
U		||||˙	
V		||||:	
W		||||:	
X		||||:	
Y		▦	
Z		▦˙	

*Mathematics
for Elementary
School Teachers*
pp. 87, 88, 99

EXPLORATION 2.8   Different Bases

The primary purpose of this exploration is to deepen your understanding of base ten—that is, your understanding of base, place value, and the role of zero. The structures of a system are often best seen by putting the familiar in an unfamiliar context; this was the reason for the Alphabitian exploration. Many elementary teachers have their students explore different bases. Some of these teachers have told me that the explorations with different bases result in their students coming to better understand place value and how base ten works and that this knowledge translates to better problem solving in base ten.

Working in different bases is often difficult at first. If you find yourself groping about, consider the following tools to help you:

Connections?

Can I connect what I know about base ten to this new base?

Problem-Solving Strategies

What if I made a table? Would that help?

As I make my tables, what patterns do I see?

Might it help to stop and say what the symbols mean?

PART 1: Learning different bases

Base six

The following activities are designed to help you understand base six.

1. **a.** Take some time, individually or together as a group, to learn how to count in base six.

    b. Describe what helped you to accomplish this task.

2. Try the following exercises, both to assess your understanding and to stretch your understanding. Answers are at the end of this section.

    What comes after 25_{six}? after 555_{six}? after 1235_{six}?

    What comes before 40_{six}? 300_{six}? 12340_{six}?

3. Describe one area in which you had initial difficulty, for example, what comes after 555? Describe the difficulty and what helped you get unstuck.

Base two

The following activities are designed to help you understand base two, which is the base for computers (0 = off and 1 = on).

1. **a.** Take some time, individually or together as a group, to learn how to count in base two.

    b. Describe what helped you to accomplish this task.

2. Try the following exercises, both to assess your understanding and to stretch your understanding. Answers are at the end of this section.

    What comes after 110_{two}? after 1011_{two}? after $101,111_{two}$?

    What comes before 100_{two}? 111_{two}? 1010_{two}?

3. Describe one area in which you had initial difficulty, for example, what comes after 1011? Describe the difficulty and what helped you get unstuck.

Base sixteen

While base two is the base for computers, many people have a hard time working with a base where the numbers change so quickly, and so computer numbers are often written in base sixteen, which requires new symbols because you don't get to one-zero in base sixteen until you have passed all the single digits in base ten. This is the conventional numeration for the symbols in base sixteen: 1, 2, 3, 4, 5, 6, 7, 8, 9, a, b, c, d, e, f, 10. That is, the symbol *a* represents the amount we would call 10; *b* represents the amount we would call 11, and so on.

1. **a.** Take some time, individually or together as a group, to learn how to count in base sixteen.

    b. Describe what helped you to accomplish this task.

2. Try the following exercises, both to assess your understanding and to stretch your understanding. Answers are at the end of this section.

    What comes after 39_{sixteen}? after $7f_{\text{sixteen}}$? after abc_{sixteen}? $78f_{\text{sixteen}}$?

    What comes before 100_{sixteen}? 450_{sixteen} 1010_{sixteen}?

3. Describe one area in which you had initial difficulty, for example, what comes after 78f? Describe the difficulty and what helped you get unstuck.

PART 2: Similarities and differences among the bases

At this point, you can count in many different bases. In fact, if you stop to think for a moment, you could count in base three without further instruction, as long as you understand what it means to say base three.

1. Take some time to look at the different bases. What patterns do you see in all bases? What patterns do you see in some bases but not other bases? Also note what sparked this discovery—perhaps looking at a table or hearing a comment by someone else.

2. Listen to the patterns observed by other students in the class. In the space below, note new patterns that you heard about.

PART 3: Translating from another base into base ten

It is important to note that the goal of this part is not to "get" the "procedure" for translating from one base to another but rather: (1) to give you practice applying the problem-solving tools you are developing, and (2) to deepen your understanding of base ten, which is the base that you will "teach" to your future students.

Think of meeting people who operated in the bases we have just studied.

1. Each person would say that *they* have base "ten." Explain why each of the bases is really base ten, that is, base one-zero.

2. **a.** If the person using base two says that there are 1101 people from her country, how many would that be in base ten?

    b. If the person using base five says that there are 34 from her country, what is that in base ten?

    c. If the person using base six says that there are 23 from her country, what is that in base ten?

    d. If the person using base sixteen says that there are 18 from her country, what is that in base ten?

3. Let's do the same thing, but with larger amounts.

    a. If the person using base two says that there are 100110 people from her country, how many would that be in base ten?

b. If the person using base five says that there are 403 from her country, what is that in base ten?

c. If the person using base six says that there are 203 from her country, what is that in base ten?

d. If the person using base sixteen says that there are 34a from her country, what is that in base ten?

4. Think about how you translated from each of these bases into base ten. Develop an algorithm that can be used for each of the bases; that is, a procedure that would enable anyone, following your directions, to translate from any base into base ten.

PART 4: Translating from base ten into other bases

Now, let us look at the process for translating from base ten into other bases.

1. Let's say each of these people arrived here 34 (base ten) days ago. Translate this amount into each of the bases.

2. Let's say each of these people will be staying here for 208 more (base ten) days. Translate this amount into each of the bases.

3. Think about how you translated from base ten into each of these bases. Develop an algorithm that can be used for each of the bases; that is, a procedure that would enable anyone, following your directions, to translate from base ten into that base.

4. **a.** Describe the most important learnings from translating to and from base ten. In each case, first describe what it was that you learned and then describe how you learned it.

   b. Describe something that you are still not clear about.

Answers to Exploration 2.8 questions

Part 1: Base six

30_{six} comes after 25_{six}. 1000_{six} comes after 555_{six}. 1240_{six} comes after 1235_{six}.
35_{six} comes before 40_{six}. 255_{six} comes before 300_{six}. 12335_{six} comes before 12340_{six}.

Part 1: Base two

111_{two} comes after 110_{two}. 1100_{two} comes after 1011_{two}. 110000_{two} comes after 101111_{two}.
11_{two} comes before 100_{two}. 110_{two} comes before 111_{two}. 1001_{two} comes before 1010_{two}.

Part 1: Base sixteen

$3a_{sixteen}$ comes after $39_{sixteen}$. $80_{sixteen}$ comes after $7f_{sixteen}$. $abd_{sixteen}$ comes after $abc_{sixteen}$.
$790_{sixteen}$ comes after $78f_{sixteen}$.

$ff_{sixteen}$ comes before $100_{sixteen}$. $44f_{sixteen}$ comes before $450_{sixteen}$. $100f_{sixteen}$ comes before $1010_{sixteen}$.

EXPLORATION 2.9   A Place Value Game

Materials

- A six-sided die in which the 6 is covered with a 0.
- The game boards provided. The grids have spaces for a 4-digit number, a 3-digit number, a 2-digit number, and a 1-digit number.

Rules for competitive version

1. Each player has a separate game board.
2. After each roll of the die, players decide where on the grid to write the number.
3. After ten rolls, each player determines the sum of the four numbers—that is, the 4-digit, 3-digit, 2-digit, and 1-digit numbers. The person with the largest sum wins.

1. Play the first game as follows:

   a. Do not speak during the game.

   b. In the blanks provided on the game board, record the order of the ten throws of your die. For example, 5 3 4 0 2 4 5 2 3 4.

   c. After the first game, take some time to think about your choices. Also think about whether you learned something from playing the game or from watching the other player. Can you express what you learned in a way that would make sense to someone who has not played the game? Write down the strategies that you learned from the first game.

2. Play the second game as follows:

   a. Record the order of the ten throws of your die.

   b. Every time you write a number in the grid, explain why you chose that location.

   c. After the game, reflect on your strategy. Describe the strategies you learned from playing the second game or from watching the other player.

3. Summarize your strategies for this game, and briefly justify them.

4. What did you learn from playing this game?

Game boards for EXPLORATION 2.9

1. __ __ __ __ __ __ __ __ __ __   2. __ __ __ __ __ __ __ __ __ __

Sum: _____   Sum: _____

EXPLORATION 2.10   How Big Is Big?

In understanding the structure of our base ten numeration system, it is also important to develop number sense, something that will be emphasized throughout the book. One aspect of number sense is having a "feel" for the relative size of large numbers. In this exploration, you will develop a sense of million and billion.

1. When I was younger, McDonald's would update the number under the arches that told how many hamburgers they had sold: over 100 million, over 200 million, and so on. Once they got past 1 billion, they stopped updating. For years the sign has simply said "over billions sold."

    a. If we laid 1 million McDonald's hamburgers in a line, how long would the line be? Pick a unit that you can sense. For example, if the answer were 500,000 inches, you would need to pick a larger unit, because no one I know can say what in their life is about 500,000 inches, but most people can say what is about 8 miles, which is equivalent to 500,000 inches.

    b. How long would 1 billion hamburgers be?

2. **a.** If we made a stack of 1 million pennies high, how tall would the stack be?

    b. If we made a stack of 1 billion pennies high, how tall would the stack be?

3. **a.** How much would 1 million pennies weigh?

    b. How much would 1 billion pennies weigh?

4. *On the Day You Were Born* is a book that appeals to almost all young children. As they get older, they can explore these questions, which you can do now!

    a. How many days old are you?

    b. How many seconds have passed since you were born?

    c. How old will you be a million seconds from now?

    d. How old will you be a billion seconds from now?

5. **a.** About how many blades of grass are on a football field?

    b. How much space would 1 million blades of grass take up?

    c. One billion blades?

    d. One trillion blades?

6. Pick an object of your choice and show how much 1 million of that object would be.

CHAPTER 3

The Four Fundamental Operations of Arithmetic

I n Chapter 2, you explored three fundamental concepts of the elementary school curriculum: sets, function, and numeration. This chapter builds on that knowledge so that you can see how our numeration system is connected to the fundamental operations of addition, subtraction, multiplication, and division. Understanding these connections will help you develop "operation sense," which in turn will help you with mental arithmetic—doing computations in your head—and estimation.

EXPLORATION 3.1 Computation in Alphabitia

In Chapter 2 you put yourself in the role of a member of the Alphabitian tribe (Exploration 2.7), and in that role you created a numeration system. You will be asked to remain Alphabitians for a while longer and to add, subtract, multiply, and divide in that system.

PART 1: Addition

1. Make up and solve at least two addition story problems in Alphabitia, and then discuss in your group how you solved them. Respond to the following questions:

    a. What knowledge of counting and place value did you use?

    b. If you encountered difficulty, what knowledge or tool(s) got you past the difficulty?

2. **a.** Now try these Alphabitian problems:

$$
\begin{array}{r} BB \\ +CC \\ \hline \end{array}
\qquad
\begin{array}{r} BC \\ +CD \\ \hline \end{array}
$$

   b. What knowledge of counting and place value did you use?

   c. If you encountered difficulty, what knowledge or tool(s) got you past the difficulty?

3. Now let us examine some addition problems with larger numbers. Focus on making sense of the process, not simply on "getting" the procedure.

    Do the following problems according to your instructor's directions.

$$
\begin{array}{r} BCD \\ + DB \\ \hline \end{array}
\qquad
\begin{array}{r} B\,0B \\ +ABC \\ \hline \end{array}
\qquad
\begin{array}{r} C0\,D \\ +CAB \\ \hline \end{array}
\qquad
\begin{array}{r} D0D0 \\ +BABA \\ \hline \end{array}
$$

4.   *Finding missing digits*

Determine missing digits to make this problem work: B_D + C_ = CBB.

PART 2: Subtraction

1.   Make up two subtraction story problems and solve them. Solve them in a way that makes sense to you—by using manipulatives, diagrams, or numerals. Have at least one of the problems involve regrouping.

2.   **a.** What previous knowledge (about numeration systems, place value, addition, and so on) did you need to solve the problem?

b. If you encountered difficulty, what knowledge or tool(s) got you past the difficulty?

3.   Take some time to solve those subtraction problems below.

The goal is to be able to determine the difference in a way that makes sense to you; there is no right way to solve these problems. Check each answer before moving on.

a. DB −AD	**b.** D0 −BB	**c.** CBD −ADC	**d.** DCB −ABD
e. C0 0 −BAB	**f.** D00 −B0B	**g.** C0 B −BAD	**h.** A0 0 − CA
i. B0 A − BD	**j.** B0 AC − BCD	**k.** A0B0C − B0CA	**l.** A0B00 − D0B

PART 3: Multiplication

1.   Make up a multiplication table for Alphabitia.

2.   Use this multiplication table as needed to solve the following problems.

a. AB × C	**b.** BB × AC	**c.** CAB × BC	**d.** B0A × CB	**e.** B0 0C × D0B

PART 4: Division

1.   Solve these division problems.

a. B)DB   **b.** C)ABA   **c.** B)DCA   **d.** C)BBBB   **e.** BC)B0DC

SECTION 3.1   Exploring Addition

Can you remember learning to add and subtract in first and second grade? As adults you probably don't think twice about adding and subtracting, but these operations are often difficult for many youngsters, especially when regrouping is involved. One way to help you better understand what these operations mean is to have you do problems in a different setting.

Mathematics
for Elementary
School Teachers
p. 117

EXPLORATION 3.2   Mental Addition

1.   Do in your head the seven computations below. Briefly note the strategies that you used, and try to give names to them.

   One mental tool that all students have is being able to visualize the standard algorithm in their heads. For example, for part (a), you could say, "$9 + 7 = 16$, carry the 1, then $5 + 3 = 8 +$ the carried 1 makes 9; the answer is 96." However, because you already own that method, I ask you not to use it here but to try others instead.

2.   Share your strategies in your small group. Note any strategies that you heard that you did not use but would like to use.

3.   In your group, select two or three strategies to describe to the class. Make up a name for each strategy.

4.   After hearing the class presentations, write down the strategies that you like best.

	Sum	What you did	Name
a.	39 + 57		
b.	78 + 25		
c.	46 + 19		
d.	625 + 147		
e.	790 + 234		
f.	68 + 55		
g.	295 + 398		

Mathematics
for Elementary
School Teachers
pp. 118, 120, 123

EXPLORATION 3.3   **Addition: Children's Algorithms and Alternative Algorithms**

PART 1: Children's algorithms

Below are descriptions of several solution paths for the problem 48 + 26—paths that children commonly invent when they are not shown how to add but, rather, are provided with rich questions to develop their understanding of place value and then are asked to figure out these problems on their own. In each case,

a. Make up and solve additional problems until you understand how the algorithm works.

b. Determine whether that solution path will work for larger numbers: for 385 + 562, for 476 + 508, and for 3245 + 683.

c. Write your first draft of why the algorithm works.

1. Add up.

   The child says: 48 + 20 = 68; 68 + 6 = 74

2. Add each place, and then combine.

   The child says: 40 + 20 is 60, 8 + 6 = 14, 60 + 14 = 74

3. Compensate.

   The child says: take 2 from 26 and give it to 48; 50 + 24 = 74

PART 2: Alternative algorithms

Below are several algorithms that have enjoyed popularity in different countries at different times. In each case,

a. Make up and solve more problems until you are confident that you understand how the algorithm works.

b. Write your first draft of why the algorithm works.

c. Compare the algorithm to the standard addition algorithm in the text with respect to the five criteria Hyman Bass described, which are given on page 142 of the text.

Note that most people are biased toward the algorithm they grew up with. For these algorithms, answer the questions thinking as a person who grew up with this algorithm.

Cross out

Add from left to right. When you see the needed regrouping in the place to the right, cross out and increase the current place by 1.

        358   3 + 5 = 8; look to the right—because 5 + 7 > 9, put a 9 in the hundreds place.
      + 574   5 + 7 = 12; look to the right—because 8 + 4 > 9, put a 3 in the tens place.
        8̶2̶2   8 + 4 = 12; put a 2 in the ones place.
        93

The "Scratch" or "Adding Up" Algorithm

At the right, you can see how 564 + 378 is computed with the "scratch" or "adding up" algorithm. You start with the ones place and find the sum of each place. If the sum is less than 10, you simply record the sum at the top. If the sum is 10 or greater, you put the digit from the ones place of the sum above and then cross out the number in the place to the left and increase it by 1. This algorithm is quite probably the origin of the phrase *adding up*.

        94
        8̶3̶2
        564
        378

The Partial Sums Algorithm

The partial sums algorithm was developed in India over 1000 years ago.

```
 369
 478
  17
 13
  7
 847
```

The Lattice Algorithm

The lattice algorithm, also popular in the Middle Ages, is one that many of my students like.

```
+ 4  8  5
  2  4  9
 0╱1╱1╱
 ╱6╱2╱4
 7  3  4
```

First, you find the sum of the digits in each place and put each of the sums in the box, as shown. To determine the answer, you extend the diagonal line segment inside each box and then add the numbers that are in the same "chute," as shown above.

Mathematics
for Elementary
School Teachers
p. 130

EXPLORATION 3.4   Addition and Number Sense

You will be asked to grapple with some different kinds of problems for each of the operations. The purpose is to develop number and operation sense that can be seen as common sense with numbers and operations.

In each problem, be prepared to explain your justification for your answers.

1. I'm thinking of 4 numbers whose sum is greater than 100.

    Tell whether each of the following must be true, might be true, or can't be true. If you are not sure, try to find a counterexample.

    a. All four numbers are greater than 20.

    b. If two of the numbers are less than 25, the other two must be greater than 25.

    c. All four numbers are two-digit numbers.

2. Determine the value of a, b, and c that will make this addition problem true:

$$
\begin{array}{r}
abc \\
+cba \\
\hline
625
\end{array}
$$

3. Determine which of these answers are reasonable. You have about 5 seconds to make your determination. That is, make your determination, using number sense, without doing any pencil-and-paper work or trying to add the numbers mentally

$$
\begin{array}{r}
40 \\
38 \\
42 \\
+45 \\
39 \\
48 \\
42 \\
\hline
340
\end{array}
\qquad
\begin{array}{r}
65,482,588 \\
+\ 47,546,716 \\
\hline
103,546,224
\end{array}
\qquad
\begin{array}{r}
999 \\
+\ 9,999 \\
899 \\
\hline
10,627
\end{array}
$$

SECTION 3.2   Exploring Subtraction

Mathematics for Elementary School Teachers p. 137

EXPLORATION 3.5   Mental Subtraction

Most people find it more difficult to determine exact answers to subtraction problems in their heads. However, if you think about the various models for subtraction that we have discussed, there are many possibilities: take-away, comparison, adding up, and the fact that the difference between two numbers tells us how far apart they are on the number line, to name but a few.

1.   Do in your head the seven computations shown below. Briefly note the strategies that you used, other than simply doing the standard algorithm in your head. (Once again, this strategy is not bad; it's just one that everybody already has.) Try to give names to your strategies.

2.   Share your strategies in your small group. Note any strategies that you heard that you did not use but would like to use.

3.   In your group, select two or three strategies to describe to the class. Make up a name for each strategy.

4.   After hearing the class presentations, write down the strategies that you like best.

Difference	What you did	Name
a. $\begin{array}{r} 65 \\ -28 \end{array}$		
b. $\begin{array}{r} 71 \\ -39 \end{array}$		
c. $\begin{array}{r} 80 \\ -36 \end{array}$		
d. $\begin{array}{r} 324 \\ -275 \end{array}$		
e. $\begin{array}{r} 1000 \\ -\ 378 \end{array}$		
f. $\begin{array}{r} 82 \\ -38 \end{array}$		
g. $\begin{array}{r} 71 \\ -32 \end{array}$		

*Mathematics
for Elementary
School Teachers*
pp. 138, 139, 141

EXPLORATION 3.6   ## Subtraction: Children's Algorithms and Alternative Algorithms

PART 1: Children's algorithms

Below are descriptions of several solution paths for the problem 92 − 38—paths that children commonly invent when they are not shown how to subtract but, rather, are provided with rich questions to develop their understanding of place value and then are asked to figure out these problems on their own.

In each case,

a. Make up and solve additional problems until you understand how the algorithm works.

b. Determine whether it will work for larger numbers: for 837 − 375, for 904 − 268, and for 800 − 358.

c. Write your first draft of why the algorithm works.

1.   Subtract down.

The child says: 92 − 30 = 62; 62 − 8 = 54

2.   Subtract the tens, and then do the ones separately.

90 − 30 = 60; 60 − 8 = 52; 52 + 2 = 54

3.   Subtract in each place; it's ok if you go negative.

$$
\begin{array}{ll}
92 & 90 - 30 = 60 \\
\underline{-\ 38} & 2 - 8 = -6 \\
60 & 60 - 6 = 54 \\
\underline{-\ 6} & \\
54 &
\end{array}
$$

4.   Compensate: Add 2 to each number.

92 − 38 becomes 94 − 40 = 54

PART 2: Alternative algorithms

Below are several algorithms that have enjoyed popularity in different countries at different times. In each case,

a. Make up and solve more problems until you understand how the algorithm works.

b. Write your first draft of why the algorithm works.

c. Compare the algorithm to the standard addition algorithm in the text with respect to the five criteria Hyman Bass described, which are given on page 142 of the text.

The Indian Algorithm

This is one of the earliest algorithms for subtraction and was popular in India almost a thousand years ago. It is called the *reverse method* and has similarities to the standard algorithm used in the United States in that it involves a "borrowing" step and a "payback" step. We begin at the left and subtract the digits in the farthest left place. We proceed place by place. As long as no regrouping is required, we just sail along, as shown in Step 1. In this example, when we get to the ones place, we encounter the problem of not being able to subtract 4 from 2. As in the algorithm commonly used, we place a 1 above the 2 that is in the ones place, giving that place a value of 12. Now we do the subtraction $(12 - 4)$ and put the difference of 8 below. Then comes the payback: We cross out the 2 in the tens place of the answer and replace it with a 1. The correct answer is now 418.

Step 1	*Step 2*	*Step 3*
632	$\overset{1}{632}$	$\overset{1}{632}$
-214	-214	-214
42	42	$4\cancel{2}8$

A Multicultural Algorithm

This algorithm appears in the first arithmetic book published in Europe (in 1478 in Treviso, Italy). In the second edition of this book, I referred to this as the Treviso algorithm. Then I had a student who grew up in Japan. As we were exploring addition, she showed me how she learned to subtract, and the procedure she was taught is identical to the Treviso algorithm.

Here is how she described what she does: To subtract $93 - 68$, because $8 > 3$ you look at what number is needed to add to 8 to make it 10, which is 2. You add that 2 to the 3 and get 5. Then you must decrease the 9 by 1 to get 8, and $8 - 6 = 2$, so the difference is 25.

In the Treviso book, the authors used the word *complement*. That is, the complement of 1 is 9, the complement of 2 is 8, and so on. Whenever you can't subtract the bottom number from the top number (and remain positive), you add the complement of the bottom number to the top number.

93	$8 + 2 = 10$	$\overset{8}{\cancel{9}}3$
$-\underline{68}$	$2 + 3 = 5$	$-\underline{68}$
	Place 5 in the ones place	25
	Decrease 9 by 1 and subtract	

632	$4 + 6 = 10$	$6\overset{2}{\cancel{3}}2$
$-\underline{214}$	$6 + 2 = 8$	$-\underline{214}$
	Place 8 in the ones place	418
	Decrease 3 by 1 and subtract	

*Mathematics
for Elementary
School Teachers*
p. 143

EXPLORATION 3.7 **Subtraction and Number Sense**

1. Fill in the blanks to make this subtraction problem

$$
\begin{array}{r}
4\ 0\ 6\ 8 \\
-\ \square\square\square\square \\
\hline
2\ 1\ 4\ 9
\end{array}
$$

2. Determine the values of a and b that will make this subtraction problem true:

$$
\begin{array}{r}
6aa \\
-1b8 \\
\hline
47b
\end{array}
$$

3. Determine which of these answers is reasonable. You have about 5 seconds to make your determination. That is, make your determination, using number sense, without doing any pencil-and-paper work or trying to subtract the numbers mentally.

$$
\begin{array}{r}
82{,}418 \\
-64{,}269 \\
\hline
17{,}149
\end{array}
\qquad
\begin{array}{r}
268{,}414 \\
-145{,}684 \\
\hline
152{,}860
\end{array}
$$

SECTION 3.3 Exploring Multiplication

The concepts of, and algorithms for, multiplication and division are more complex than those for addition and subtraction. It is crucial that your understanding of multiplication and division be richer than simply knowing how to multiply and how to divide.

Mathematics for Elementary School Teachers p. 152

EXPLORATION 3.8 Patterns in the Multiplication Table

PART 1: Base ten

1. Examine the multiplication table and describe the patterns that you see.

	1	2	3	4	5	6	7	8	9	10	11	12
1	1	2	3	4	5	6	7	8	9	10	11	12
2	2	4	6	8	10	12	14	16	18	20	22	24
3	3	6	9	12	15	18	21	24	27	30	33	36
4	4	8	12	16	20	24	28	32	36	40	44	48
5	5	10	15	20	25	30	35	40	45	50	55	60
6	6	12	18	24	30	36	42	48	54	60	66	72
7	7	14	21	28	35	42	49	56	63	70	77	84
8	8	16	24	32	40	48	56	64	72	80	88	96
9	9	18	27	36	45	54	63	72	81	90	99	108
10	10	20	30	40	50	60	70	80	90	100	110	120
11	11	22	33	44	55	66	77	88	99	110	121	132
12	12	24	36	48	60	72	84	96	108	120	132	144

2. Select one pattern that your group will share with the class.

   a. Describe this pattern as though you were talking on the phone to a friend who missed class today but who has a multiplication table handy. The purpose of your description is simply to get the person to see the pattern.

   b. Exchange descriptions with another group. If they easily interpreted your description, great. If not, revise your description (like a second draft of an essay). Discuss the parts of your description that were not clear to the other group. That is, what was unclear about the description, and why do you think the revised wording is better?

   c. What did you learn from describing a pattern and reading another group's description?

3. Now explain *why* the pattern occurs.

PART 2: Circle clocks

Circle clocks (also called *star patterns*) have been used by many teachers to introduce many different mathematical concepts and to provide a visual connection for these ideas. The following steps use these circle clocks to give you another perspective on the basic multiplication facts.

Take out the Base Ten Circle Clocks on page 61. You will draw on one clock for each of the rows in the multiplication table from 2 through 9. The directions are to look only at the units digit. For example, if we look at the multiplication facts in row 2 and write down only the units digit, we have 2, 4, 6, 8, 0, and then the numbers repeat. Start your pencil at 2 on the circle; then draw a line from 2 to 4, then from 4 to 6, and so on.

1. Complete the circle clock for the multiples of 2, and describe the pattern as though you were talking to someone on the phone.

2. Complete the circle clocks for each of the other multiples.

3. What similarities do you see among the different circle clocks? Can you explain why those similarities occur?

4. Predict the shape of the circle clock for the multiples of 11. Explain your reasoning. Then draw the pattern. If your prediction was correct, great. If it wasn't, or if you weren't able to make a prediction, either describe the knowledge that you weren't able to apply or describe what enables you to understand why the pattern is what it is.

PART 3: Other bases

Make a set of multiplication tables in different bases determined by your class or your instructor.

1. **a.** Describe patterns that seem to be true in all bases.

   b. Describe patterns that are true in some but not all bases.

2. Select one pattern and describe the pattern, as though on the phone to someone who has the tables in front of him or her.

3. Now describe the *why* behind the pattern.

Base Ten Circle Clocks for EXPLORATION 3.8, PART 2

Counting by 1
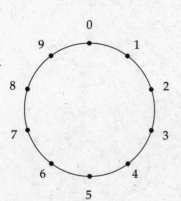

Counting by 2
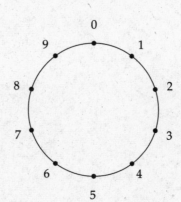

Counting by 3
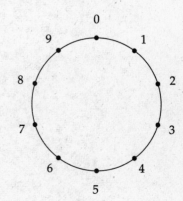

Counting by 4
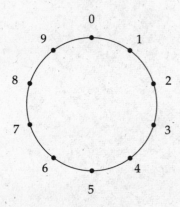

Counting by 5
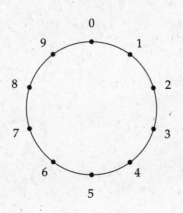

Counting by 6
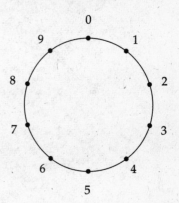

Counting by 7
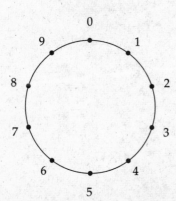

Counting by 8
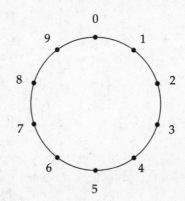

Counting by 9
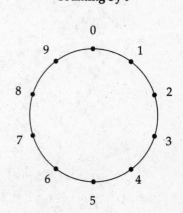

Mathematics
for Elementary
School Teachers
p. 154

EXPLORATION 3.9   **Mental Multiplication**

1.   Do in your head the seven computations shown below. Briefly note the strategies that you used, other than simply doing the standard algorithm in your head. Try to give a name to each strategy.

2.   Share your strategies in your small group. Note any strategies that you heard that you did not use but would like to use.

3.   In your group, select two or three strategies to describe to the class. Make up a name for each strategy.

Product	What you did	Name
a. 82 × 5		
b. 24 × 25		
c. 23 × 12		
d. 638 × 2		
e. 70 × 40		
f. 25 × 14		
g. 45 × 19		

EXPLORATION 3.10	Differences Between Multiplication and Addition*

1. Write a story problem for 16×18.

2. Now make 16×18 with base ten blocks and with unifix cubes. Think about different ways for each manipulative and think about colors with the unifix cubes.

3. Compare your constructions with students in other groups. What do you notice?

4. One mental strategy for addition is called *compensation*.

   For $16 + 18$, it works like this: $16 + 18 = 14 + 20$. That is, take 2 from 16 and give it to 18. If we were to use this strategy for 16×18 we would say that $16 \times 18 = 14 \times 20$. Is this true? If so, why does this strategy work for both operations? If not, why does it work for addition but not for multiplication?

5. Another mental strategy for addition has us add each place separately and then add the two results.

   For $16 + 18$, it works like this: $16 + 18 = (10 + 10) + (6 + 8)$—that is, add the places separately. If we were to use this strategy for 16×18, we would say that $16 \times 18 = (10 \times 10) + (6 \times 8)$. Is this true? If so, why does this strategy work for both operations. If not, why does it work for addition but not for multiplication?

6. Another mental strategy for some addition problems is to round both addends to the next "nice" number and then take away the differences.

   For $16 + 18$, it works like this: $16 + 18 = 20 + 20 - 4 - 2$. If we were to use this strategy for 16×18, we would say that $16 \times 18 = 20 \times 20 - 4 - 2$. Is this true? If so, why does this strategy work for both operations? If not, why does it work for addition but not for multiplication?

7. Answer each of the following questions after some careful thinking and discussion with other students.

   a. What makes multiplication work differently from addition?

   b. What do you need to think about multiplication as you modify strategies that are appropriate for addition to make them appropriate for multiplication?

   c. What did you learn from this exploration?

*Adapted from *Developing Mathematical Ideas: Numbers and Operations, Part 1 Building a System of Tens. Facilitator's Guide* by Deborah Schifter, Virginia Bastable, and Susan Jo Russell. Copyright © 1999 by the Education Development Center, Inc. Published by Pearson Education, Inc., publishing as Dale Seymour Publications, an imprint of Pearson Learning Group. Used by permission.

EXPLORATION 3.11 Cluster or String Problems

Before children learn the standard algorithm for multiplication, it is helpful to have them figure on their own how to determine products of larger amounts. They need to string together smaller problems, which requires them to use the distributive principle in meaningful ways. In each of the problems below, you will build the answers to multidigit problems using only your knowledge of the basic multiplication facts and your understanding of place value.

In each case, explain how you can use the results of previous computation(s) or multiplication facts to find the result of the current computation. One string for 14×8 follows.

$1 \times 8 = 8$	multiplication fact
$10 \times 8 = 80$	multiply by 10 means you add a 0
$4 \times 8 = 32$	multiplication fact
$14 \times 8 = 112$	14 eights = 10 eights + 4 eights

1. 2×4
    3×4
    20×4
    23×4

2. 4×6
    40×6
    2×6
    200×6
    1×6
    241×6

3. 40×3
    5×3
    45×3
    45×30
    45×29

4. 5×6
    50×6
    25×6
    25×12
    250×12
    251×12

Find another string to make 45×29.   Find another string to make 251×12.

5. Make up your own strings to solve the following problems. You can use: your multiplication facts up through 10×10, the properties of multiplication, and your knowledge of addition, subtraction, and place value. Justify each step, as in the problems above.

    a. 56×6 **b.** 372×6 **c.** 72×15 **d.** 34×18

    e. 71×32 **f.** 45×36 **g.** 123×45 **h.** 206×52

Mathematics for Elementary School Teachers p. 157

EXPLORATION 3.12   Understanding the Standard Multiplication Algorithm

When I taught fourth grade every day in 2003, I was surprised at how hard it was for the fourth-graders to master multidigit multiplication. Therefore, I went back and explored this process more. For the purpose of this exploration, assume that you do not have an algorithm for multiplying multidigit numbers.

1. The problem 16 × 7 is shown in two ways.

   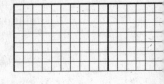

   a. How does that bold vertical line make it easier to find the answer to 16 × 7?

   b. Explain how you found the answer to 16 × 7.

2. Now using the Base Ten Graph Paper in the Appendix, cut out and find the answers to the following problems. Explain how you found the answer.

a. 17	b. 14	c. 23	d. 27
× 8	× 6	× 7	× 8

3. At the right are three 23 × 14 rectangles. One first step in finding a more efficient way to find the product is to use our base ten knowledge to break this problem into 6 small rectangles (10 × 10, 10 × 10, 3 × 10, 4 × 10, 4 × 10, and 3 × 4). Write the six products in the diagram and add them.

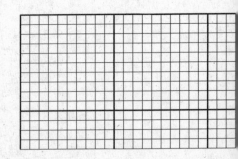

4. A next step is to decompose the problem into only two rectangles.

   a. Use a colored pencil or pen to draw a vertical line so that you have 20 × 14 and 3 × 14.

   b. Explain why the sum of these two problems is equal to 14 × 23.

   c. Multiply 14 × 23 longhand. What connections do you see between the two partial products in b. and what we just did?

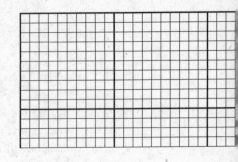

5. a. Draw a horizontal line so that you have 23 × 10 and 23 × 4.

   b. Explain why the sum of these two problems is equal to 23 × 14.

   c. Multiply 23 × 14 longhand. What connections do you see between the partial products and what we just did?

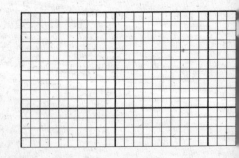

6. Now combine the two cuts above to make four smaller problems out of 23 × 14. Write those four problems. Look at the expanded-form representation of 23 × 14, and look at your picture. Describe the connections you see.

$$\begin{array}{r} 20 + 3 \\ \times \quad 10 + 4 \\ \hline 80 + 12 \\ 200 + 30 \end{array}$$

*Mathematics
for Elementary
School Teachers
p. 159*

EXPLORATION 3.13 Alternative Algorithms for Multiplication

"Before the introduction of the Arabic notation, multiplication was difficult, and the division even of integers called into play the highest mathematical faculties. Probably nothing in the modern world could have more astonished a Greek mathematician than to hear that, under the influence of compulsory education, the whole population of Western Europe . . . could perform the operation of division for the largest numbers. This fact would have seemed . . . a sheer impossibility."[1]

Egyptian "Duplation"

How do you think people multiplied before base ten? Imagine finding 35×28 with the Egyptian numerals we saw in Section 2.3 of the text—that is, ∩∩∩IIIII times ∩∩IIIIIIIII. The Egyptians, who did not have a numeration system with a base, developed an algorithm for multiplying based on the idea of multiplication as repeated addition and using doubling. This was the most practical way to multiply in the Western world before base ten was adopted.

The Egyptians would have done the following to multiply 35×28:

∩∩∩IIIII — 35
∩∩∩∩∩∩∩ — 70
Ϩ∩∩∩∩ — 140
ϨϨ∩∩∩∩∩∩∩∩ — 280
ϨϨϨϨϨ∩∩∩∩∩ — 560

You might be able to figure out the Egyptian method more easily if you use Hindu-Arabic numerals.

35
70
140
280
560

The Egyptians would then have added

Ϩ∩∩∩∩ — 140
ϨϨ∩∩∩∩∩∩∩∩ — 280
ϨϨϨϨϨ∩∩∩∩∩ — +560
————
980

to come up with
ϨϨϨϨϨϨϨϨϨϨ∩∩∩∩∩∩∩∩

1. a. Do some more problems on your own until you feel confident using the Egyptian duplation algorithm.

   b. Write directions for using this algorithm and give them to a friend. See whether the friend can get the correct answer on the basis of your directions. If she or he can, great. If not, have a conversation and find where the directions "went wrong." Keep revising the directions until your directions make sense to your friend. If possible, try the new directions on another friend.

   c. Now that you know how this algorithm works, try to explain the *why* of each step, as was done for the standard algorithms in the textbook.

The Lattice Algorithm

Just as there was a lattice algorithm for addition, there is also one for multiplication. See whether you can figure out how it works for 45×28. Just as in the standard algorithm, there are four partial products in this algorithm. One of my grandmothers told me that this is how she learned multiplication when she was a little girl.

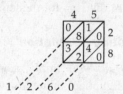

2. **a.** Do some more problems on your own until you feel confident using this algorithm. Note that this and the next algorithm can be applied to larger problems, such as multiplying two-digit by three-digit numbers or three-digit by three-digit numbers.

    b. As before, write directions for using this algorithm and give them to a friend. Have your friend multiply two numbers using your directions.

    c. As before, justify each step of this algorithm.

    d. Can you adapt this algorithm for three-digit by two-digit problems? For three-digit by three-digit problems?

The "Cross Product" or "Lightning" Algorithm

This algorithm first appears in the Treviso Arithmetic. It became one of the more popular algorithms for multiplying. The direct translation of the Treviso instructions for multiplying 56 by 48 is given below.

"(M)ultiply 6 times 8, making 48; write 8 under the units and reserve 4. Then multiply crosswise, thus: 4 times 6 are 24, and 5 times 8 are 40; add 24 and 40, giving 64, and add the 4 which was carried making 68; write 8 and carry 6. Now multiply the tens by the tens, thus: 4 times 5 are 20, and 6 to carry are 26, which is written in its proper place. The result is therefore 2688. We then say that 48 times 56 are 2688. In this same way you can perform all other cross multiplications."[2]

$$
\begin{array}{r}
56 \\
\times\ \underline{48} \\
2688
\end{array}
$$

3. **a.** Do some more problems on your own until you feel confident using this algorithm.

    b. As before, write directions for using this algorithm and give them to a friend. Have your friend multiply two numbers using your directions.

    c. As before, justify each step of this algorithm. *Hint:* Express the numbers in expanded form, such as 56 as $50 + 6$.

    d. Can you adapt this algorithm for three-digit by two-digit problems? For three-digit by three-digit problems?

EXPLORATION 3.14 Multiplication and Number Sense

1. Without doing any mental or pencil-and-paper computing, determine whether this statement is true or false.

$$246 \times 8 = (123 \times 4) + (123 \times 4)$$

2. This problem is similar to puzzles archaeologists have to solve; the circles represent places where they could not read the numbers. We have recovered these fragments of this multiplication problem. What was the problem and what is the answer?

$$
\begin{array}{r}
\bigcirc 4 \\
\times \quad 3\bigcirc \\
\hline
57\bigcirc \\
\bigcirc\bigcirc\bigcirc \\
\hline
\bigcirc\bigcirc\bigcirc 6
\end{array}
$$

3. **a.** The result of 37×35 will be in the hundreds, thousands, or ten thousands?

    b. The result of 736×35 will be in the hundreds, thousands, or ten thousands? Explain your choice.

4. One way to estimate is to round one number and then we can get the estimate more easily. What if we were doing 23×82? We could get close by doing 20×82 or 23×80. Both will be too low, but which will be closest to the actual number? Again, answer this question without doing anything on paper. Explain your choice.

5. Make a multiplication problem where the product is between 3000 and 3500.

6. Is this answer reasonable? Why or why not? You have about 5 seconds to make your determination. That is, make your determination, using number sense, without doing any pencil-and-paper work or trying to multiply the numbers mentally.

$$
\begin{array}{r}
294 \\
\times \quad 62 \\
\hline
19{,}968
\end{array}
$$

7. The object is to place the digits 1, 2, 3, 4, and 5 in the boxes in such a way as to obtain the greatest product.

    a. Make your first guess and briefly describe your reasoning.

    b. Make another guess and describe your reasoning.

    c. Continue this process until you think you have obtained the greatest product.

SECTION 3.4   Exploring Division

*Mathematics
for Elementary
School Teachers*
p. 170

EXPLORATION 3.15   Different Models of Division

1.   Make up a story problem for 15 ÷ 3. Assume that you are a young child who does not yet know division. Describe how you would determine the answer. You may use pictures to describe your solution path, but you are not required to.

2.   Compare your stories with those of other members of your group. Just as we found that there are different models for addition, subtraction, and multiplication, there are different models for division.

   a. Sort your stories into two or more groups. Explain the common characteristic(s) of the stories in each group.

   b. Give a name to each of the groups and briefly justify that name.

   c. After listening to the class discussion, name and briefly define each of the models for division.

3.   Consider the problem 92 ÷ 4. Make up a story problem for these numbers that fits each of your models. Still assuming that you don't know the formal idea of division, describe how you would determine the answer.

EXPLORATION 3.16   Understanding Division Computation

Using your understanding of what division means, solve the following problems by a means other than using your standard algorithm.

1. a. $169 \div 14$
    b. $576 \div 8$
    c. $4352 \div 35$

*Mathematics
for Elementary
School Teachers*
p. 174

EXPLORATION 3.17 Mental Division

1. Do in your head the five computations shown below. Briefly note the strategies that you used, and try to give names to them.

2. Share your strategies in your small group. Note any strategies that you heard that you did not use but would like to use.

3. In your group, select two or three strategies to describe to the class. Make up a name for each strategy.

4. After hearing the class presentations, write down the strategies that you like best.

Quotient	What you did	Name
a. $800 \div 40$		
b. $125 \div 5$		
c. $432 \div 8$		
d. $1463 \div 7$		
e. $174 \div 6$		

EXPLORATION 3.18   Dealing with Remainders

Real-life problems involving division generally require not just the ability to do the computation but also the ability to interpret the result. This exploration enables you to explore the issue of remainders more deeply.

1. The Allentown Elementary School is going to the Science Museum in a nearby city. There are 369 students going, and each school bus can hold 24 students. How many school buses will be needed? Do this problem by yourself and then discuss your answer with members of your group.

2. Make up a story problem involving 31 divided by 4 in which the answer is

    a.   7   **b.**   8

    c. $7\frac{3}{4}$ **d.**   7 remainder 3

3. After listening to other students' story problems, what did you learn about division from this exploration?

Mathematics
for Elementary
School Teachers
p. 177

EXPLORATION 3.19   The Scaffolding Algorithm

Many students have trouble with long division. Common problems include poor multiplication facts, not being able to remember the sequence, and misplacing the digits in the quotient. The standard algorithm can be unforgiving, as the example at the right shows. After repeated failure with the standard algorithm, many students simply give up; many of my students in this course have told me of bad experiences learning the division algorithm.

$$
\begin{array}{r}
69 \ \text{R } 24 \\
8\overline{)576} \\
\underline{48} \\
96 \\
\underline{72} \\
24
\end{array}
$$

An alternative algorithm called the scaffolding algorithm has been helpful in giving students a sense of success, which builds their confidence. It also develops their ability to use guess–check–revise and reinforces their multiplication skills.

Here is how it might work with the problem 576 ÷ 8. The student is asked to estimate the quotient.

Let's say the student's first guess is 60. The student multiplies 60 × 8 to see how many 8s have been "used up" and how many are left to be dealt with. The 60 is placed at the top. (*Note:* The diagram for each step is below.)

The student is now asked the same question with the remaining 96. Let's say the student guesses 10. The 10 is placed above the 60 (that is, in the "answer" space), and the student takes away the 10 groups of 8, or 80.

We now have 16 to deal with. The student sees that 8 × 2 = 16 and places the 2 in the answer space. The student now adds 60 + 10 + 2 to get the answer.

At the far right is another example of this algorithm; note that the actual numbers in the answer space and the number of steps depend on the student's guesses.

$$
\begin{array}{cccc}
 & & & 1 \\
 & & 2 & 6 \\
 & 10 & 10 & 20 \\
60 & 60 & 60 & 50 \\
8\overline{)576} & 8\overline{)576} & 8\overline{)576} & 64\overline{)4953} \\
\underline{480} & \underline{480} & \underline{480} & \underline{3200} \\
96 & 96 & 96 & 1753 \\
 & \underline{80} & \underline{80} & \underline{1280} \\
 & 16 & 16 & 473 \\
 & \underline{16} & \underline{16} & \underline{384} \\
 & 0 & 0 & 89 \\
 & & & \underline{64} \\
 & & & 25
\end{array}
$$

1. **a.** Do some more problems on your own until you feel confident using the scaffolding algorithm.

    b. Write directions for using this algorithm and give them to a friend. See whether the friend can divide on the basis of your directions. If he or she can, great. If not, have a conversation and find where the directions "went wrong." Keep revising the directions until your directions make sense to your friend. If possible, try the new directions on another friend.

    c. Now that you know how this algorithm works, try to explain the *why* of each step, as was done for the standard algorithm in the textbook.

    d. What advantages might this algorithm offer in teaching?

EXPLORATION 3.20 Understanding the Standard Algorithm

1. Mathematical power comes from knowing what each step of an algorithm means. The middle column in the table below describes the verbalization of each step in the pencil-and-paper algorithm. The right column asks you to explain the why for each step. A story for this problem is: A janitor has moved all 72 desks out of 3 classrooms. How many desks go into each classroom?

$\begin{array}{r} 24 \\ 3\overline{)72} \\ 6 \\ \overline{12} \\ \underline{12} \end{array}$	What we do	Why we do it
	a. 3 "gazinta" 7 two times	**a.** What is going on mathematically when we say "gazinta"?
	b. Place the 2 above the 7.	**b.** Why do we place the 2 above the 7?
	c. 3 × 2 = 6. Place the 6 below the 7.	**c.** Why do we multiply 3 × 2?
	d. Subtract 6 from 7.	**d.** Why do we subtract?
	e. Bring down the 2.	**e.** What is going on mathematically when we "bring down" the 2?
	f. 3 "gazinta" 12 four times. Place the 4 above the 2. Next, 3 × 4 = 12. Place the 12 below the other 12.	**f.** Why do we repeat these three steps again: gazinta, multiply, subtract?

2. Many students have trouble when there are zeros in the quotient. Explain these steps in the long-division problem below.

$\begin{array}{r} 304 \\ 27\overline{)8208} \\ 81 \\ \overline{108} \\ \underline{108} \end{array}$	Algorithm	Justification
	a. 27 doesn't go into 8.	**a.** What is going on mathematically here?
	b. 27 goes into 82 three times.	**b.** What is going on mathematically here?
	c. 27 × 3 = 81; put the 3 above the 2.	**c.** How do you know where to put the 3?
	d. Subtract 82 − 81 and bring down the 0.	**d.** Why do we bring down the 0, since we are bringing down nothing?
	e. Since 27 doesn't go into 10, put a 0 above the 0.	**e.** Why?
	f. Bring down the 8. 27 "gazinta" 108 four times. . .	**f.** Why?

*Mathematics
for Elementary
School Teachers*
p. 181

EXPLORATION 3.21   Division and Number Sense

1.   Find the divisor.

$$\begin{array}{r} 65R \\ \hline ?\overline{)614} \end{array} \qquad \begin{array}{r} 71R5 \\ \hline ?\overline{)644} \end{array}$$

2.   Find the quotient.

$$\begin{array}{r} 85R17 \\ \hline 56\overline{)?} \end{array}$$

3.   Is this answer reasonable? Why or why not? You have about 5 seconds to make your determination. That is, make your determination, using number sense, without doing any pencil-and-paper work or trying to multiply the numbers mentally.

$$\begin{array}{r} 704 \\ \hline 55\overline{)40000} \end{array}$$

4.   Determine which of these is the better estimate for $\frac{935}{22}$ with computing all three quotients:

$$\frac{1000}{25} \quad \text{or} \quad \frac{1000}{20}$$

5.   One way to estimate division problems is to round both numbers up or both numbers down. What if we were estimating $\frac{225}{9}$? Which would be the better estimate: $\frac{250}{10}$ or $\frac{230}{10}$? Why?

Looking Back

As you look back on the explorations with addition, subtraction, multiplication, and division, stop and reflect on what you have learned.

1.   Briefly describe at least three important learnings. Below are four questions to start the reflection process.

  • Did you learn more about what one or more of the operations mean?

  • Do you better understand how one or more of the algorithms work?

  • Do you better understand how two or more of the operations are connected?

  • Did working through these explorations help you to understand place value more deeply or to understand how zero works in base ten?

2.   Select and discuss one of these learnings. Your discussion needs to contain these three aspects:

a. A description of *what* you learned

b. Within this description, an explanation of the "why" behind the "what"

c. An explanation of why this learning seems important to you

EXPLORATION 3.22   Developing Operation Sense

PART 1: Determining the appropriate operation

In each of the problems below, select the correct operations by a means other than random trial and error. Briefly explain the reasoning behind your guesses.

1. $2 \bigcirc 4 \bigcirc 3 = 11$
2. $43 \bigcirc 24 \bigcirc 68 = 1100$
3. $684 \bigcirc 418 \bigcirc 942 \bigcirc 246 = 962$
4. $(624 \bigcirc 319) \bigcirc (722 \bigcirc 699) = 41$

PART 2: Algebra and operation sense

In each of the cases below, the actual numbers have been replaced by variables. Write the number sentence that would correctly determine the answer.

1.   Germaine bought A CDs for B dollars each. He sold each CD for C dollars. How much profit did he make?

2.   Gary bought A boxes of golf balls each containing B balls. If he sold each box for C dollars, how much money did he make?

3.   Mandy wants to buy a car for A dollars. She has currently saved B dollars. If she has C months to save up for the car, on average she needs to save D dollars each month.

4.   Tasha is going on a trip of A miles. If her car gets about B miles per gallon and gas will cost about C dollars per gallon, how much can she expect to spend on gas?

PART 3: Which answer is reasonable?

Determine which of these answers are reasonable. You have about 5 seconds to make your determination. That is, make your determination, using number sense, without doing any pencil-and-paper work.

1.   There were 35 students in a class and each student paid $28 for a field trip. The total cost was about $600.

2.   There are 648 students in a school that has made a deal to buy each student a computer for $415 per student. The total cost of the program will be about $18,000.

3.   The Boosters Club has to pay for the banquet. The meal cost $2143, the trophies cost $436, the senior gifts cost $562, and the raffle prizes cost $82. The club had $8265 in the bank. After the banquet, the balance is about $5000.

4.   A craftsperson has made 352 wooden spoons for the holiday season, and he will distribute these to 12 dealers. Each dealer will get about 25 spoons.

EXPLORATION 3.23   Operation Sense in Games

Greatest amount

Materials

- One die with the following numbers: 1, 2, 3, 4, 5, 6.

Directions for playing the game

- Roll the die four times and record the numbers.
- The object is to make the value of the given expression as great as possible.

1. $\square \times \square + \square - \square$

    a. Before the first game, describe a master strategy that you think will work for all cases.

    b. Play several games.

    c. Write down your master strategy if it has changed.

2. $\square \times \square \div \square + \square$

    a. Do you think the same master strategy you developed in Step 1 will apply, or do you need to modify it?

    b. Play several rounds, recording your numbers and your reasoning.

    c. Write down your master strategy if it has changed.

3. $\dfrac{(\square \times \square)}{(\square - \square)}$

    a. Before rolling the die, describe a master strategy for this game.

    b. Play several rounds, recording your numbers and your reasoning.

    c. Write down your master strategy if it has changed.

4. Play any of the games above. This time the goal is to make the value of the given expression as small as possible.

Target

This game is adapted from the *Fifth-Grade Book* in the NCTM Addenda Series (p. 48).

Materials

- Three dice:

    Die 1: 1, 2, 3, 4, 5, 6

    Die 2: 1, 2, 3, 4, 5, 6

    Die 3: 7, 8, 9, 10, 11, 12

Directions for playing the game

- Pick a target number.
- Roll the three dice.
- Use the three numbers you roll and any combination of operations to get as close as possible to the target number you have chosen.

1. For each target number and three numbers, write your answer and your reasoning.

EXPLORATION 3.24 How Many Stars?

Without counting each star, determine how many stars are on this page. Describe and justify your process.

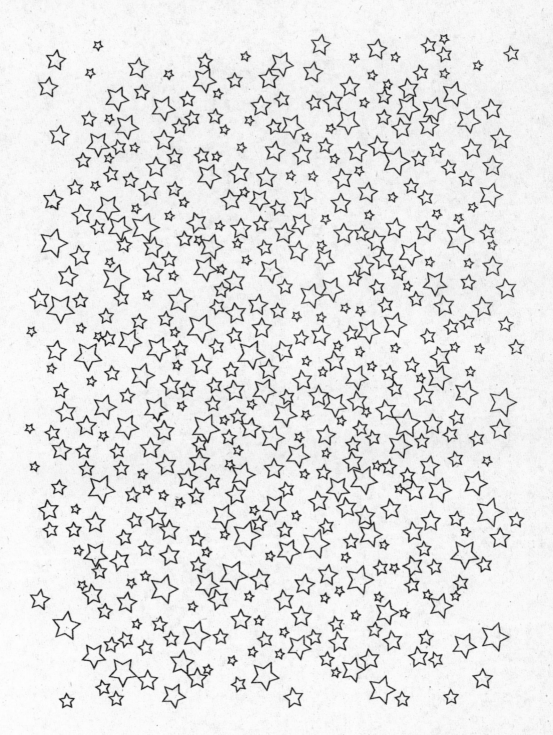

CHAPTER 4

Number Theory

Many people do not realize that mathematical concepts and ideas often have visual representations. For example, in Chapter 3 you investigated different visual representations of multiplication. In this chapter's explorations, you will be able to visualize some of the concepts that we will study in the text. These explorations can be done with elementary school children, yet they offer interesting challenges to the adult student too.

In each of the following explorations, you will not only lay the foundation for learning some important number theory ideas but will also have the opportunity to develop more mathematical tools—making and testing predictions. Students often tell me that when they think of mathematics, they think of numbers and computations. However, making and testing predictions has been crucial not only to the development of mathematics but also to the development of civilization. Young children are constantly making and testing predictions; I have cleaned up many messes my children made as a result of erroneous predictions! People make and test predictions regularly in their personal and professional lives. Yet somehow, this important human activity seems to be absent from much of school mathematics.

SECTION 4.1   Exploring Divisibility and Related Concepts

As you discovered in Chapter 3, being able to decompose and recompose a number in different ways is crucial to understanding computation algorithms for the four operations and to doing mental arithmetic and estimation confidently. Depending on the situation, we might decompose 48 as $40 + 8$, as $50 - 2$, as $8 \cdot 6$, or as $2 \cdot 2 \cdot 2 \cdot 2 \cdot 3$. In the following exploration, as you play the Taxman game more times, you will discover that one kind of decomposition of a number will enable you to increase your score, and you will see new relationships among numbers.

EXPLORATION 4.1   Taxman

I don't know who invented this game, but I have seen variations in many different places. The version I present to you is one in which a team of two (or more) players competes against the "taxman."

Materials

- Array of consecutive numbers (see the game sheets on page 83).

Rules

1. The team crosses out a number.
2. The taxman crosses out all the proper factors of that number.
3. The team crosses out another number. A number can be crossed out only if at least one of its proper factors has not yet been crossed out. For example, you can't cross out 14 if 1, 2, and 7 have all been crossed out already. Play continues until none of the remaining numbers has a proper factor that has not been crossed out.
4. At the end of the game, the taxman gets all remaining numbers!
5. To determine your score, add up all the numbers your team crossed out.

1. Take out the Taxman Game Sheet on page 83 and play several games. As you are playing, write down observations about strategies. For example, is there a best first move? Is there a worst first move? Is it better to go for big numbers earlier or later? Listen to the class discussion about strategies, and write down anything you learned from that discussion.

2. Take out the Taxman Game Sheet for Step 2 on page 83 and play several games. As you are playing, write down observations about strategies. For example, is there a best first move? Is there a worst first move? Is it better to go for big numbers earlier or later? Listen to the class discussion about strategies, and write down anything you learned from that discussion.

3. Answer the following questions.

   a. If we were to play a game with a game sheet from 1 to 40, what would you recommend for the first move? Justify your choice.

   b. What would be the worst first move? Why?

   c. Let's say a team played the game with numbers 1–30 and their score was 234. To find the taxman's score, they could add up all of the taxman's numbers, but there is an easier way to deduce the taxman's score. Can you find it?

4. As you have seen, some numbers have many factors and some have very few. The ancient Greeks were fascinated by relationships among numbers, and they came up with three terms to classify numbers with respect to their factors. If you add the sum of the proper factors of a number, that sum might be greater than, equal to, or less than the number itself. In the former case, a number is said to be abundant, in the second case perfect, and in the third case deficient.

   a. Classify the first 30 numbers in this manner.

   b. Write down any observations from your work.

   c. Predict what group the following numbers would be in, and briefly describe your prediction: 47, 48, 49, 50

   d. Predict a number above 50 that you think is likely to be deficient, and explain your choice.

   e. Predict a number above 50 that you think is likely to be abundant, and explain your choice.

   f. How does this work connect to the game or help you with strategies for it?

Taxman Game Sheet for EXPLORATION 4.1

| 1 | 2 | 3 | 4 | 5 | 6 | 7 | 8 | 9 | 10 | Your score ____ |

| 11 | 12 | 13 | 14 | 15 | 16 | 17 | 18 | 19 | 20 | Taxman ____ |

| 1 | 2 | 3 | 4 | 5 | 6 | 7 | 8 | 9 | 10 | Your score ____ |

| 11 | 12 | 13 | 14 | 15 | 16 | 17 | 18 | 19 | 20 | Taxman ____ |

| 1 | 2 | 3 | 4 | 5 | 6 | 7 | 8 | 9 | 10 | Your score ____ |

| 11 | 12 | 13 | 14 | 15 | 16 | 17 | 18 | 19 | 20 | Taxman ____ |

Taxman Game Sheet for EXPLORATION 4.1, Step 2

1	2	3	4	5	6	7	8	9	10	
11	12	13	14	15	16	17	18	19	20	
21	22	23	24	25	26	27	28	29	30	Score ____

1	2	3	4	5	6	7	8	9	10	
11	12	13	14	15	16	17	18	19	20	
21	22	23	24	25	26	27	28	29	30	Score ____

1	2	3	4	5	6	7	8	9	10	
11	12	13	14	15	16	17	18	19	20	
21	22	23	24	25	26	27	28	29	30	Score ____

Exploring Prime and Composite Numbers

*Mathematics
for Elementary
School Teachers*
p. 196

EXPLORATION 4.2   Factors

An exploration many elementary teachers use to help students develop number sense is to challenge them to make as many different rectangles as they can using a certain number of unit squares. For example, how many different rectangles can be made with six squares?

We can make two different rectangles with six unit squares. If we represent them by their length and width, we have a 6×1 rectangle and a 3×2 rectangle. Using mathematical terminology, we can also say that 6 has these four factors: 1, 2, 3, 6.

When elementary school children investigate all the different rectangles they can make for each number, the exploration not only reinforces their multiplication facts but also addresses other important areas: problem solving, communication, reasoning, and making connections. Because you are able to think at a more abstract level, we will modify the instructions given to the children.

1. Take out the Factors Table on page 87. In the table, list the factors of each of the first 25 natural numbers.

2. Describe any observations, hypotheses, and questions you have as a result of filling out and looking at the table.

3. One of the themes of this book is the power of different representations.

   a. Take out the Number of Factors Table on page 89. Use the data in your Factors Table from Step 1 to complete the table.

   b. Note any observations, hypotheses, and questions that you have as a result of filling out and looking at this table.

   c. In one sense, the numbers in each column of the table constitute a "family." Can you give a name to any of the families? If so, write down this name (it does not have to be a "mathematical" name) and describe the characteristics that are common to all members of the family.

   d. Suppose we were to continue to look at the factors of numbers beyond 25. Predict (and briefly describe your reasoning) the next number that will have 2 factors, the next number that will have 3 factors, and the next number that will have 4 factors.

   e. Now extend the table until you have the next number that has 2 factors, 3 factors, and 4 factors. Describe any insights or discoveries from the extension.

   f. Can you predict the first number that will have 7 factors? What will it look like? Justify your reasoning.

   g. There are other possible families of numbers, based on the number of factors. For example, if we combine the 3-factor families, the 5-factor families, and the 7-factor families into a large family called "odd number of factors," how might you describe the characteristics that are common to all members of *this* family?

Factors Table for EXPLORATION 4.2, Step 1

Number	Factors							
1	1							
2	1	2						
3	1	3						
4								
5								
6								
7								
8								
9								
10								
11								
12								
13								
14								
15								
16								
17								
18								
19								
20								
21								
22								
23								
24								
25								

Number of Factors Table for EXPLORATION 4.2, Step 3

1 factor	2 factors	3 factors	4 factors	5 factors	6 factors	7 factors	8 factors	9 factors
1	2	4	6	16	12		24	
	3	9	8		18			
	5	25	10		20			
	7		14					
	11		15					
	13		21					
	17		22					
	19							
	23							

largest Prime 97

*Mathematics
for Elementary
School Teachers*
p. 214

EXPLORATION 4.3   **Finding All Factors of a Number**

In this exploration, we will be working with patterns (recognizing, describing, extending), making predictions and hypotheses, and using several related problem-solving tools, including being systematic and making tables.

1. Determine all the factors of 36. Show your work. Describe how confident you are that you have found *all* the factors and why you are very confident, somewhat confident, or not very confident.

2. Select some other numbers, or use ones given by your instructor, and find all of their factors. If you feel that your method is efficient, move on. If you don't feel that your method is very efficient, take some time to stop and think about how else you might determine all the factors of a number.

3. Describe your present method for finding all the factors of a number, as though you were talking to a classmate who missed this exploration.

4. Describe any other observations or hypotheses you have made up to this point.

5. Now compare your methods with those of your partner(s). If you like a method that you heard about, summarize this new method in your own words.

6. How could you use the factors of 36 that you found in Step 1 to find all the factors of 72 without starting all over?

7. The number 200 has 12 factors: 1, 2, 4, 5, 8, 10, 20, 25, 40, 50, 100, and 200. This number 200 can be rewritten as $200 = 2^3 \cdot 5^2$. This representation is an example of a *prime factorization,* which is discussed in Section 4.2 of the text. Can you see any way to determine all 12 factors of 200 just from seeing this prime factorization?

8. Find all the factors of 100. Now use this information to find all the factors of 300. Explain your thinking.

SECTION 4.3   Exploring Greatest Common Factor and Least Common Multiple

The following explorations, like the previous ones, actually involve many concepts. As with the previous explorations in this chapter, your ability to decompose numbers and then analyze those decompositions and see relationships will enable you to answer the questions.

EXPLORATION 4.4   African Sand Drawings

Elaborate sand drawings are part of the storytelling of the Chokwe people who live in parts of Angola and Zaire. The figures, called *lusona* (the plural is sona), are used in telling stories that are part of the culture. Critical to all these drawings is that they can be made without lifting the finger; that is, they are continuous line drawings. You can find more about these drawings from *Geometry From Africa: Mathematical and Educational Explorations*, by Paulus Gerdes; from *Africa Counts*, by Claudia Zaslavsky; and from the Web.

We will investigate one specific kind of lusona and determine which sona can be traced continuously and which cannot. First, we create a rectangular array of dots. We then trace a path in this rectangular array with certain rules. This process is illustrated below. You can begin anywhere, but you must move in a diagonal direction. When you get to the outside boundary of the array, you have to turn 90 degrees. (The behavior of the line is identical to a billiard ball hitting the edge of the table.) This first example is known as the antelope because, with a few artistic additions, the figure can be made to resemble an antelope, shown at the right.

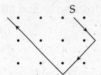 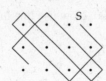 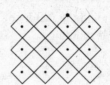 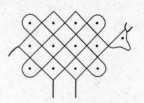

In this first example, when you get back to the starting point, you have now enclosed, or captured, all of the dots in the original rectangular array. We have also captured all the dots in one *continuous* path. However, not all rectangular arrays can be traced in one continuous path. In the rectangular array below, we come back to the starting point without having finished the drawing. That is, it requires two paths to enclose all the dots. (I have shown the second path with a heavier line.) This will be true for this array no matter where you start.

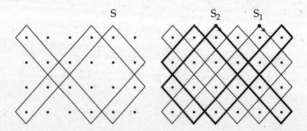

Instructions

1.   Experiment with rectangular arrays of different sizes until you can predict the number of turns it will take to complete a lusona of any size. For example, it took one turn to complete a 4 × 3 array, and it took two turns to complete a 4 × 6 array.

2.   State your answer and how you arrived at your answer.

Here are some tips that previous students have found helpful so that their tracings of the paths are valid. The example of a 4 × 6 will be shown here.

1. Lightly trace a rectangle around the borders of this array. After becoming familiar with the process, many students find they don't need the border and can do it with the imaginary border, like the Chokwe do.

2. Begin at a point on the rectangle that is directly above a dot.

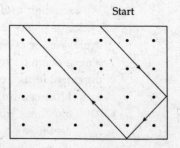

3. The "bounce" when you hit the border is always at a right angle, like a ball bouncing off the border of a pool table.

4. Draw the path with a ruler. The edge of the ruler will always be parallel to the line of dots on the diagonal direction.

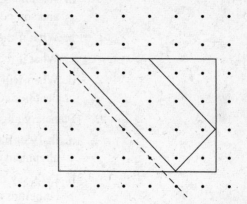

5. When you get back to the starting point, if you haven't "captured" all the dots, then pick another point, and trace a path that begins with that point.

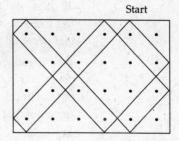

6. Repeat this process until all the dots are captured. In this case, it requires two paths to complete the drawing. We can symbolize this by writing $n(4, 6) = 2$. Since the 3 × 4 array could be traced in one path, we can write $n(3, 4) = 1$.

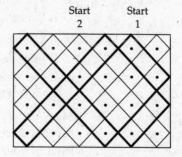

Mathematics
for Elementary
School Teachers
pp. 211, 220

EXPLORATION 4.5 Cycles

We encounter cycles all the time—weather cycles, biological cycles, economic cycles, and lif
cycles of plants and animals. The elementary and middle schools in Keene all follow a 6-da
cycle. We see cycles in many words, such as *tricycle* and *bicycle*, and the words *circle* and *cycl*
have a common root. Below are several problems that involve cycles. Solving them draws o
important ideas from number theory.

1.   Most carnivals and amusement parks have a Ferris wheel ride. Some larger ones have dou
     ble Ferris wheels, and others have large and small Ferris wheel rides. Suppose two sister
     decide to go on the big Ferris wheel, which makes one complete rotation in 50 seconds. A
     younger brother is afraid of the large wheel and goes on the small Ferris wheel with hi
     father; the small Ferris wheel makes one complete rotation in 20 seconds.

     a. Assuming that both rides begin at the same time, when will the two sisters and the fathe
        and son be at the bottom again?

     b. What if the periods of the Ferris wheels were 45 seconds and 25 seconds?

     c. What if the periods were 80 seconds and 50 seconds?

     d. When will the two sisters and the father and son all be at the top at the same time witl
        the 50- and 20-second Ferris wheels?

2.   Planetary alignments

     a. In 1988 Earth, Jupiter, Saturn, and Uranus were all in alignment. When will they be i
        alignment again if the orbits of the three other planets are 12, 30, and 84 Earth years?

     b. Let's say that Earth, Mercury, and Venus are aligned at a certain time. How long will i
        be before all three are again aligned, if the yearly cycles of Mercury and Venus are 8
        and 224 Earth days?

3.   Our physical cycle completes one life span (or cycle) in 23 days. The emotional cycle last
     28 days, and the intellectual cycle lasts 33 days. These cycles are hypothesized to rise an
     fall like sine waves in trigonometry. Let's say a person's physical, emotional, and intellec
     tual cycle all peaked. When would be the next time all three would peak?

4.   Your organization is having a fund raiser and will be selling hot dogs. You anticipate sellin
     about 200 hot dogs. You go to the store and find that the hot dogs come in packages of 12
     but the hot dog buns come in packages of 8. What number around 200 would give you th
     same number of hot dogs and hot dog buns?

5.   An interactive museum has many programs that cycle throughout the day. Assume that al
     of the programs begin at 9 A.M.

     a. If you want to see two programs that have, respectively, 10- and 15-minute cycles, wha
        is the earliest you could be finished?

     b. If you want to see three programs that have, respectively, 10-, 15-, and 20-minute cycles
        what is the earliest you could be finished?

     c. If you want to see three programs that have, respectively, 10-, 12-, and 15-minute cycles
        what is the earliest you could be finished?

6.   Many children have had fun with this one. The directions are to write your name on graph
     paper and see the patterns that occur with different lengths. You are finished when the las
     letter of your name appears on the last column, thus making a rectangle. The case for a 3
     letter name is shown at the top of the next page for three different lengths.

4 columns				5 columns					7 columns						
P	A	T	P	P	A	T	P	A	P	A	T	P	A	T	P
A	T	P	A	T	P	A	T	P	A	T	P	A	T	P	A
T	P	A	T	A	T	P	A	T	T	P	A	T	P	A	T

a. Using a sheet of "Other Base Graph Paper" at the end of this book, try your name with various lengths. Describe any patterns you see, your observations, and any questions.

Two questions that often come from the initial exploration are described below. Exploring them involves some wonderful opportunities to understand various aspects of number theory, as well as to develop your competence with the five process standards stated by NCTM. Read the two questions and then follow the directions immediately below the questions.

b. *Question 1*   Background: You will have noticed that in some cases, there is a diagonal pattern going from bottom left to top right (as in the 4- and 7-column example) above, sometimes there is a diagonal pattern going the other way (as in the 5-column example), and sometimes there is no real diagonal pattern.

The question: Can you predict when a rectangle will have the first diagonal pattern, the second diagonal pattern, or no diagonal pattern, given the length of letters in the "name" and the length of the column?

c. *Question 2*   Background: We can also look at the dimensions of the rectangle formed by one complete cycle. In the case of a 3-letter name, the rectangle for a 4-column pattern is 4×3, the rectangle for a 5-column pattern is 5×3, and the rectangle for a 7-column pattern is 7×3.

The question: Can you predict the dimensions of the rectangle, given the length of letters in the "name" and the length of the column? Note that there are many possibilities explaining the makeup of the finished rectangle—odd numbers, even numbers, prime numbers, numbers with common factors, and many more!

CHAPTER 5

Extending the Number System

M any beginning elementary teachers think that they will never have to teach fractions, integers, or decimals unless they teach upper elementary school. This is not true. Even in kindergarten, children encounter the idea of fractions in terms of dividing. For example, "We each get half." Second graders will often ask, "Why can't you take 8 from 2?" when a teacher introduces regrouping (for example, $52 - 38$). Long before formal work with decimal operations, children encounter decimals when they deal with money. Therefore, it is essential that all elementary teachers have an understanding of the relationship of these three systems (integers, fractions, and decimals) to whole numbers and to each other.

SECTION 5.1 Exploring Integers

Although most students do not encounter integers until middle school, many elementary school children are aware of them, and quite a few will be able to work with them. Therefore, it is important that the elementary teacher be comfortable with integers from a conceptual point of view— that is, know how they connect to positive whole numbers and how operations with integers are like and unlike operations with positive whole numbers. The following explorations are designed to help you better understand integer operations.

*Mathematics
for Elementary
School Teachers*
p. 238

EXPLORATION 5.1   Understanding Integer Addition

When we examined addition with positive whole numbers, one model we used to represent this operation was the set (discrete) model. Let us use this model to develop rules for integer addition. We will use black dots to represent positive numbers and white dots to represent negative numbers.[1] Thus the figure below is a representation of 3 + 4 in the context of addition as joining. That is, we have two sets, one containing 3 dots and the other containing 4 dots. When we combine the two sets, we have 7 dots; that is, 3 + 4 = 7.

1. Model the problem 5 + (−2). Describe and justify your process so that a reader who is not familiar with integer addition could understand both how you got the answer and why your process works.

2. Model the problem −7 + 5. Describe and justify your process so that a reader who is not familiar with integer addition could understand both how you got the answer and why your process works.

3. Model the problem −3 + (−6). Describe and justify your process so that a reader who is not familiar with integer addition could understand both how you got the answer and why your process works.

4. We have now examined all four cases that can occur when we are adding integers: both numbers positive, the larger magnitude positive, the larger magnitude negative, and both numbers negative. On the basis of your work above, can you describe a general rule for adding integers that will work for all cases?

EXPLORATION 5.2   Understanding Integer Subtraction

As you discovered in Chapter 3, diagrams of subtraction look very different with different models. We will examine integer subtraction first in the context of the take-away model of subtraction and then in the context of the comparison model of subtraction.

Subtraction as Take-away

A representation of $5 - 2$ from the take-away perspective follows. That is, the figure at the left below shows 5 objects. The figure at the right shows us taking 2 away, and thus 3 objects remain; that is, $5 - 2 = 3$.

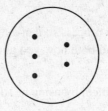 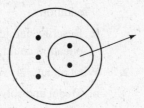

Using the "subtraction as take-away" context, model the following problems. In each case, describe and justify your process so that a reader who is not familiar with integer subtraction could understand both how you got the answer and why your process works.

a. Model the problem $2 - 5$; that is, model 2 take away 5.

b. Model the problem $-3 - 6$.

c. Model the problem $-2 - (-4)$.

d. We have now examined all possibilities (cases) that can occur when we are subtracting integers. On the basis of your work above, can you describe a general rule for subtracting integers that will work for all cases?

Subtraction as Comparison

As you saw in Chapter 3, and as most first- and second-graders will readily tell you, subtraction as comparison is quite different from subtraction as take-away. The key to being able to model integer subtraction with the "subtraction as comparison" model is to find a way to adapt our notion of subtraction as comparison so that it works with integers. For example, we could describe $6 - 2$ as comparison by saying it means "How much bigger is 6 than 2?" We can readily see that 6 is 4 bigger than 2. However, this way of describing subtraction as comparison doesn't help us with problems like $-3 - 6$, "-3 is how much bigger than 6?" Work with your partners to find a way to describe subtraction as comparison that will work for the cases we examined with take-away.

a. Model the problem $2 - 5$.

b. Model the problem $-3 - 6$.

c. Model the problem $-2 - (-4)$.

d. We have now examined all possibilities (cases) that can occur when subtracting integers. On the basis of your work above, can you describe a general rule for subtraction as comparison that will work for all four cases?

Mathematics for Elementary School Teachers
p. 243

EXPLORATION 5.3 **Understanding Integer Multiplication**

Below is a representation of $4 \cdot 3$ using chips. It uses the repeated-addition context for multiplication. That is, we have four sets of 3, or three 4 times.

1. What about $4 \cdot (-3)$?

   a. Model and solve this problem by using or adapting the idea of multiplication as repeated addition.

   b. Describe and justify your process so that a reader who is not familiar with integer multiplication could understand both how you got the answer and why your process works.

2. What about $-4 \cdot 3$?

   a. Model and solve this problem by using or adapting the idea of multiplication as repeated addition.

   b. Describe and justify your process so that a reader who is not familiar with integer multiplication could understand both how you got the answer and why your process works.

3. Model the problem $-4 \cdot (-3)$. Describe and justify your process so that a reader who is not familiar with integer multiplication could understand both how you got the answer and why your process works.

4. We have now examined all four cases that can occur when we are multiplying integers. On the basis of your work above, can you describe a general rule for multiplying integers?

EXPLORATION 5.4   Understanding Integer Division

Repeated-Subtraction and Partitioning Models

Consider the problem $8 \div 2$. We learned in Chapter 3 that we can represent this problem using the partitioning model of division, and we can represent this problem using the repeated-subtraction model of division. For review, you may want to model the problem $8 \div 2$ with the partitioning model and with the repeated-subtraction model.

1.   **a.** Model the problem $-8 \div 2$ with the partitioning model.

   b. Model the same problem with the repeated-subtraction model.

   c. Describe and justify your process so that a reader who is not familiar with integer division could understand both how you got the answer and why your process works.

Missing-Factor Model

When we get to other cases of integer division (for example, $8 \div (-2)$ and $-8 \div (-2)$), these two models become more and more complex, and at some point, they pass into the realm of "more trouble than it's worth." Therefore, let us abandon the partitioning and repeated-subtraction models for division and look to another model for division to justify the procedures for integer division: the missing-factor model. When we use this model, we find the solution to $a \div b$ by asking, "What number times b is equal to a?"

2.   Use the missing factor model to determine the answers to the following three problems. As before, describe and justify your process so that a reader who is not familiar with integer division could understand both how you got the answer and why your process works.

$$-8 \div 2$$
$$12 \div (-3)$$
$$-20 \div (-4)$$

3.   We have now examined all four cases that can occur when dividing integers. On the basis of your work above, can you describe a general rule for dividing integers?

SECTION 5.2   Exploring Fractions and Rational Numbers

The fraction explorations here have been designed to give you an opportunity to work with the fundamental concepts related to fractions. Just as we discussed different decompositions of whole numbers in Chapters 3 and 4, a key to understanding fractions is to decompose them. That is, the notion of parts and wholes connects to our compositions and decompositions with whole numbers. We shall investigate various decompositions, which in turn will deepen your understanding of the different fraction contexts and the relationship between the numerator and the denominator.

EXPLORATION 5.5   Making Manipulatives

One of the reasons for the limited success of much work with manipulatives on fractions is that students often work with premade manipulatives. Many mathematics educators believe that children will benefit tremendously by making their own manipulatives. Such an activity not only allows the students to grapple with concepts at a deeper level but also encourages more creativity on their part.

Making the Manipulatives

Circles: Using construction paper that has been cut into circles of the same size, make a set of manipulatives. The only restriction is that you cannot use a protractor. For example, you cannot make $\frac{1}{4}$ circles by measuring 90-degree angles. In addition to physical tools, you will also need to use a combination of problem-solving strategies, including reasoning and guess–check–revise.

Squares: Using construction paper that has been cut into squares of the same size, make a set of manipulatives. The only restriction is that you cannot use a ruler to measure the divisions. For example, if your square is 6 inches on a side, you cannot use the ruler to make a mark every 2 inches. In addition to physical tools, you will also need to use a combination of problem-solving strategies, including reasoning and guess–test–revise.

After you have made your manipulatives, respond to the following questions.

1.   **a.** Describe how you made thirds as though you were talking to someone who missed class. Your description needs to have enough specificity so that the person could repeat what you did and see why you did it that way.

   b. Describe how you made fifths as though you were talking to someone who missed class.

2.   Describe at least two learnings that resulted from making your set of manipulatives.

3.   Describe any questions that you have at this point, either questions about how to make a particular fraction (such as ninths) or other questions that arose when you were making your sets.

4.   Alicia said that making $\frac{1}{5}$ was hard. Brandon said that it was easy because $\frac{1}{5}$ is "halfway between $\frac{1}{4}$ and $\frac{1}{6}$." Jamie said that Brandon's method can be used for lots of cases; for example, $\frac{1}{7}$ will be halfway between $\frac{1}{6}$ and $\frac{1}{8}$, and $\frac{1}{6}$ will be halfway between $\frac{1}{4}$ and $\frac{1}{8}$.

   a. What is your initial reaction to Brandon's statement? Do you feel that $\frac{1}{5}$ is halfway between $\frac{1}{4}$ and $\frac{1}{6}$ or not? Why?

   b. Discuss this question in your group. If you changed your mind, explain what changed your mind and justify your present position. If you didn't change your mind, but you feel you can better justify your response, write your revised justification.

5.   Leah called some fractions "prime fractions." What do you think she meant?

EXPLORATION 5.6 Sharing Brownies

A powerful way to have children explore fraction ideas is to pose problems involving sharing. Problems like these have been found in many articles in *Teaching Children Mathematics,* other journals, and elementary school textbooks. One interesting historical note is that many children's strategies mirror the ancient Egyptians' approach toward fractions, which is discussed in the text. For example, when children are asked how four children could share three brownies, a common solution path looks like the illustration below. Rather than give each person $\frac{1}{4}$ of a brownie, they draw solutions where the pieces will be as big as possible. That is, they prefer the solution in which each person gets $\frac{1}{2}$ of one brownie and $\frac{1}{4}$ of another to the solution of each person getting three $\frac{1}{4}$ brownies. Please stay within this spirit as you solve the following problems.

In each case, determine how to divide the brownies and write the amount each child will get. Do not add the amounts together. For example, in the problem above, the solution is $\frac{1}{2}+\frac{1}{4}$.

Brownies	Children
3	6
4	6
5	6
4	5
3	5
2	5

Mathematics
for Elementary
School Teachers
pp. 248, 249

EXPLORATION 5.7   Partitioning

PART 1: Halves

These problems are similar to ones in the fourth-grade book of an elementary textbook series! Because the figures are not simple polygons, they require the learner to apply various fraction ideas to answer the questions. In each case, shade $\frac{1}{2}$ of the polygon.

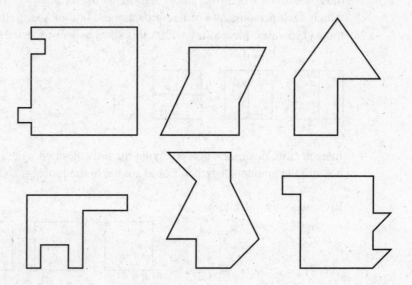

PART 2: Determining parts on a square Geoboard

These kinds of problems are also found in elementary schools and bring out the relationship between fractions and measurement. *Hint:* What do you remember about finding the area of triangles?

1.   Determine the part of the whole polygon that is shaded.

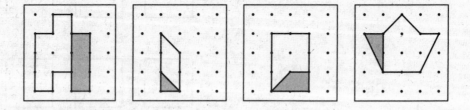

2.   What fraction of the whole Geoboard (16 squares) is represented by each of the polygons?

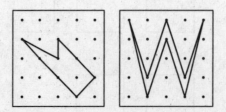

PART 3: Determining parts on a triangular Geoboard

In these problems, the unit of measurement is not a square but an equilateral triangle. Determine what fraction of the whole triangle is represented by each part.

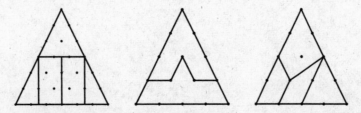

PART 4: Determining fractions of a thermometer

Look at the thermometers on page 107. In each case, determine the fraction of the thermometer that is shaded in two different ways. They are on a separate page so that you can cut them out and fold them if you wish.

PART 5: Determining fractions on a number line

Determine the value of x in problems 1–4.

1.

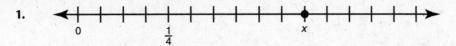

2.

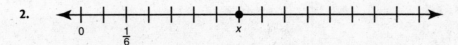

3.

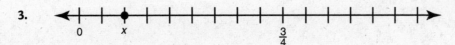

4.

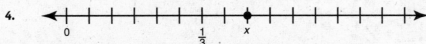

Determine the location of the designated fraction in problems 5–8.

Locate $\frac{5}{8}$.

5.

Locate $\frac{7}{15}$.

6.

Locate $\frac{5}{6}$.

7.

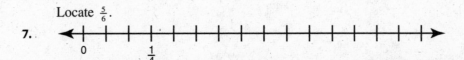

Locate $\frac{5}{6}$.

8.

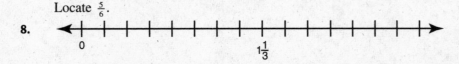

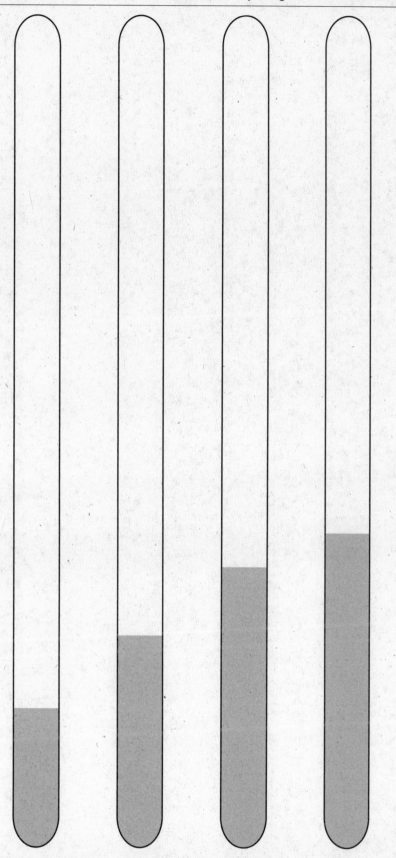

EXPLORATION 5.8 Equivalent Fractions

Your instructor will have some of the following manipulatives and concrete materials for you to use: Pattern Blocks, Cuisenaire Rods, Geoboards, circles, discrete models, number line, paper (for folding), etc.

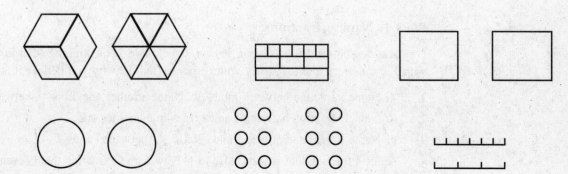

PART 1: Determining equivalence

1. Use one or more of these models to demonstrate the equivalence of $\frac{2}{3}$ and $\frac{4}{6}$.

2. Use one or more of these models to demonstrate the equivalence of $\frac{2}{3}$ and $\frac{8}{12}$.

3. Use one or more of these models to demonstrate the equivalence of $\frac{3}{4}$ and $\frac{6}{8}$.

4. Use one or more of these models to demonstrate the equivalence of $\frac{3}{4}$ and $\frac{9}{12}$.

PART 2: Extending equivalence beyond proper fractions

1. How would you demonstrate that $\frac{6}{6}$ is equivalent to 1?

2. How would you demonstrate that $3\frac{3}{8}$ is equivalent to $\frac{27}{8}$?

4. Are $\frac{3\frac{1}{2}}{7}$ and $\frac{1}{2}$ equivalent?

5. Are $\frac{0}{3}$ and $\frac{0}{8}$ equivalent? Why or why not?

*Mathematics
for Elementary
School Teachers*
pp. 254, 263

EXPLORATION 5.9   Developing Fraction Sense

After each part, compare your responses and your justification with those of your partner(s). As before, if you agree, listen to one another's justifications. If you disagree, discuss and debate until you reach agreement.

PART 1: Naming fractions[2]

For each step below, (a) give your answer, (b) explain your thinking (that is, how you came up with your answer), and (c) justify your response (that is, why you believe it is correct).

1.   **a.** Name a fraction between $\frac{1}{6}$ and $\frac{2}{6}$. **b.** Name another one. **c.** Is $\frac{1\frac{1}{2}}{6}$ between $\frac{1}{6}$ and $\frac{2}{6}$?

2.   **a.** Name a fraction between $\frac{9}{10}$ and 1. **b.** Name another one.

3.   **a.** Name a fraction between $\frac{3}{5}$ and $\frac{3}{4}$. **b.** Is $\frac{3}{4\frac{1}{2}}$ between $\frac{3}{5}$ and $\frac{3}{4}$?

4.   Name a fraction that is very close to 1. Now name a fraction that is even closer to 1.

5.   Give a value of x that makes the following statement true: $0 < \frac{4}{x} < \frac{1}{10}$.

6.   Name a fraction between 0 and $\frac{1}{10}$ that does not have a numerator of 1.

PART 2: Fraction benchmarks[3]

1.   "Wall problems" are used by some elementary teachers both as a pre-assessment and to begin a rich conversation about a topic. Do this wall problem, which is useful for beginning our work on fraction sense: Write on the wall (chalkboard or poster paper) everything you know about $\frac{2}{5}$, for example, what it looks like, how large it is, what other fractions or numbers it is close to, etc.

2.   **a.** Place each of the fractions below in one of three groups: closer to 0, closer to $\frac{1}{2}$, or closer to 1. Briefly justify your choice.

$$\frac{3}{8} \qquad \frac{2}{7} \qquad \frac{1}{3} \qquad \frac{21}{50} \qquad \frac{4}{5} \qquad \frac{7}{11} \qquad \frac{31}{181}$$

3.   What happens to the value of $\frac{5}{9}$ if

  a. the value of the numerator is increased by 1?

  b. the value of the denominator is increased by 1?

  c. the value of the denominator is decreased by 1?

  d. the value of the numerator and denominator are both increased by 2?

  e. the value of the numerator and denominator are both multiplied by 2?

  f. the value of the numerator is increased by 1 and the value of the denominator is decreased by 1?

SECTION 5.3   Exploring Operations with Fractions

As you may have already known before this course, being able to perform fraction computations and being able to apply fraction knowledge in real-life problems are not the same thing. As in Chapter 3, understanding the whys behind the whats is crucial if you are going to be an effective teacher of children. If you do the explorations in this section conscientiously, you will come out with a much stronger "fraction sense" and will better understand why the various fraction algorithms actually work!

Mathematics for Elementary School Teachers p. 262

EXPLORATION 5.10   Ordering Fractions

This kind of exploration is used in many elementary school curricula to help students develop *fraction sense* by ordering fractions. This is done *before* they have developed algorithms for equivalent fractions or converting fractions into decimals. While children will begin with manipulatives and pictorial representations, at some point we want to move beyond them, as we saw in Chapter 3 with base ten blocks. For example, demonstrating that $\frac{7}{8} > \frac{6}{7}$ cannot be done easily with a diagram, but it can be done by noting that both fractions are one piece away from 1, and since $\frac{1}{7} > \frac{1}{8}$, $\frac{6}{7}$ is missing a bigger piece, and thus $\frac{6}{7} < \frac{7}{8}$.

There are many tools that you do have, including resorting to the meaning of numerator and denominator (as used in the example above), benchmarks ($\frac{1}{2}, \frac{1}{3}, \frac{1}{4}$, etc.), and other ways of looking at fractions, for example, as the ratio of the numerator and denominator.

1.   Predict the relative value of the pairs of fractions on page 113 by inserting the symbol <, =, or > into the space between the fractions without referring to pictures, finding the least common multiple (LCM) of both fractions, or converting them to decimals. Briefly justify your choice.

2.   Order each set of fractions, again without referring to pictures, finding the LCM of both fractions, or converting them to decimals. Briefly justify your choice.

a. $\dfrac{1}{2}$ $\dfrac{2}{7}$ $\dfrac{5}{9}$

b. $\dfrac{2}{5}$ $\dfrac{3}{10}$ $\dfrac{11}{19}$

c. $\dfrac{2}{3}$ $\dfrac{13}{17}$ $\dfrac{31}{80}$

d. $\dfrac{1}{2}$ $\dfrac{1}{4}$ $\dfrac{3}{10}$ $\dfrac{5}{6}$ $\dfrac{7}{8}$

e. $\dfrac{5}{6}$ $\dfrac{2}{5}$ $\dfrac{5}{11}$ $\dfrac{8}{9}$ $\dfrac{2}{9}$

f. $\dfrac{1}{8}$ $\dfrac{2}{5}$ $\dfrac{5}{8}$ $\dfrac{5}{6}$ $\dfrac{3}{49}$ $\dfrac{3}{56}$

Table for EXPLORATION 5.10: Ordering fractions

	Fraction	>, =, or <	Fraction	Justification
a.	$\dfrac{3}{5}$		$\dfrac{3}{8}$	
b.	$\dfrac{5}{6}$		$\dfrac{7}{8}$	
c.	$\dfrac{3}{5}$		$\dfrac{5}{12}$	
d.	$\dfrac{1}{2}$		$\dfrac{17}{31}$	
e.	$\dfrac{3}{8}$		$\dfrac{2}{9}$	
f.	$\dfrac{2}{7}$		$\dfrac{3}{8}$	
g.	$\dfrac{1}{4}$		$\dfrac{2}{9}$	
h.	$\dfrac{9}{11}$		$\dfrac{7}{9}$	
i.	$\dfrac{3}{8}$		$\dfrac{2}{5}$	
j.	$\dfrac{3}{10}$		$\dfrac{9}{23}$	
k.	$\dfrac{2}{9}$		$\dfrac{3}{10}$	

Mathematics for Elementary School Teachers pp. 268, 269

EXPLORATION 5.11 Adding Fractions

While adding fractions is conceptually simpler than multiplying fractions, the algorithm for adding fractions is actually more challenging, and many children stumble with this algorithm.

1. For the moment, I would like you to suspend your knowledge that you have to get a common denominator to add fractions. Imagine that you don't have any algorithms and thus have to find the answers using the manipulatives and your knowledge of what *fraction* means. Using your manipulatives, add the following fractions and briefly explain your reasoning.

   a. $\dfrac{1}{2} + \dfrac{1}{3}$ **b.** $\dfrac{1}{4} + \dfrac{2}{3}$ **c.** $\dfrac{3}{4} + \dfrac{1}{6}$ **d.** $1\dfrac{2}{3} + 2\dfrac{3}{4}$

2. On the basis of your work above, explain why we have to find a common denominator in order to add fractions. Some students find this alternative version of the question preferable: Why can't we just add the top numbers and add the bottom numbers? For example, why isn't $\frac{3}{4} + \frac{1}{6} = \frac{4}{10}$?

*Mathematics
for Elementary
School Teachers*
p. 250

EXPLORATION 5.12   Making Sense of Wholes and Units

As you have discovered in previous explorations, it is easy to get lost in fractions in terms of distinguishing between wholes and units. For example, when finding the sum of $\frac{1}{2}$ and $\frac{1}{4}$ using a diagram, children will often give an answer of $\frac{3}{8}$. They have confused the whole and the unit. The following problems are designed to help you deepen your understanding.

PART 1: Using area models

1. **a.** If the rectangle has a value of $\frac{1}{3}$, show 1.

    b. If the rectangle has a value of $\frac{3}{4}$, show 1.

    c. If the rectangle has a value of $1\frac{1}{3}$, show 1.

2. The following questions are related to Pattern Blocks.

    a. If the hexagon has a value of $\frac{2}{3}$, show 1.

    b. If the trapezoid has a value of $\frac{3}{4}$, show 1.

    c. If two hexagons have a value of $1\frac{1}{3}$, show 1.

    d. If the trapezoid has a value of $\frac{3}{4}$, what is the value of the rhombus?

PART 2: Using discrete models

1. Draw a diagram that has the specified value. Justify your solutions.

    a. If the given diagram has a value of $\frac{2}{3}$, show 1.

    b. If the given diagram has a value of 1, show $\frac{3}{4}$.

    c. If the given diagram has a value of $2\frac{2}{3}$, show 1.

    d. If the given diagram has a value of $\frac{1}{3}$, show $\frac{1}{2}$.

2. What fraction of the circles in the diagram below are black? Your answer needs to have a denominator other than 12. Justify your answer.

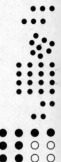

PART 3: Absent students[4]

Let's say you are teaching. It is winter, and it seems that a larger fraction of students than normal are sick. You are eating lunch in the teachers' lounge, and another teacher says, "In my class today, $\frac{2}{5}$ of the girls were absent but only $\frac{1}{5}$ of the boys were absent." What fraction of her class was absent today?

  a. Work on the problem above and show your work.

  b. What if the number of boys and girls is equal? Does this change your answer in a. or not? Explain your response.

  c. Meet with your partner(s) and discuss the problem. If you have the same answer, did you arrive at the answer in the same way? If you have different answers, listen to one another's reasoning until you can agree on one answer. Justify your answer.

  d. What is a realistic range of possible answers?

  e. What is the theoretical range of possible answers?

PART 4: How much is her share?[5]

This problem is adapted from a real-life problem. Josephine is a graduate student at Urban State College. Because her financial resources are limited, she has moved into a house with four other people. The house is heated with electricity, and the electric bill comes every two months. Josephine moved in on February 1. When the bill for January–February comes, what fraction of the bill should each person pay?

  a. Work on this problem alone and show your work.

  b. What assumptions did you make in order to solve this problem?

  c. Meet with your partner(s) and discuss the problem. If you have the same answer, did you arrive at the answer in the same way? If you have different answers, listen to one another's reasoning until you can agree on one answer. Justify your answer.

EXPLORATION 5.13   **Multiplying Fractions**

The goal of this exploration is to understand why the procedures for multiplying fractions work. Thus, as I have done in the past, I will ask you to assume you do not know the algorithm so that you can immerse yourself in the *why*; you already know the *how*!

1.   Below is a representation of $\frac{3}{4} \times 2\frac{1}{2}$. We can see this as the area of a rectangle with these dimensions. We can also see it as $\frac{3}{4}$ $2\frac{1}{2}$ times.

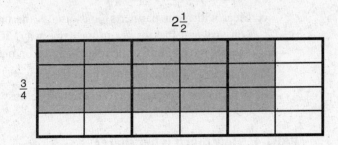

a.   In order to transform the area into a number, we need to partition the shaded area into equal-sized pieces.

   First, state how many equal-sized pieces we have. This is the numerator of our answer.

b.   Then, using your understanding of what the denominator means, determine the denominator.

c.   Discuss your solution and justification with your partners and then the whole class.

d.   What did you learn from this exploration?

2.   Let us now look at the problem $3\frac{1}{2} \times 2\frac{1}{3}$.

a.   Draw a diagram to represent this problem.

b.   Determine the answer from your diagram.

c.   Discuss your solution and justification with your partners and then the whole class.

d.   Draw the same diagram to represent the problem once again.

e.   This time partition the diagram into equal-sized pieces if your original solution did not involve partitioning the diagram into equal-sized pieces. Determine the number of pieces and determine the answer as an improper fraction.

f.   Now solve the problem using the algorithm.

g.   Compare the diagram in part (e) to the algorithm. Describe the connections you see between the diagram and the algorithm.

Mathematics
for Elementary
School Teachers
p. 278

EXPLORATION 5.14 An Alternative Algorithm for Dividing Fractions[6]

The famous Indian mathematician Brahmagupta described an algorithm for dividing fractions that is actually easier to understand, in terms of *why* it works, than the traditional invert-and-multiply algorithm. The purpose of this exploration is to help you to understand this algorithm and why it works.

For each of the problems below, do the following:

a. Represent the problem with a diagram and then solve the problem using the diagram.

b. Describe what you did in order to arrive at your answer. It is important to consider carefully what you are actually doing, because careful thinking here increases the chances of the "breakthrough" in the next step.

c. Try to connect what you did on paper to how the problem could be solved using only numbers. For example, in the first problem, the original number sentence is $4 \div \frac{2}{3}$, but regardless of the diagram you draw, you will divide each of your four units into thirds, and thus you are now solving $\frac{12}{3} \div \frac{2}{3}$.

After completing the five problems, look for commonalities in all the problems that lead to a generalization (rule) that you could use in all the division problems. That is, you are trying to go from the concrete (pictorial) level to the abstract (numerical) level so that the manipulation of the numbers will make sense, just as you found for whole-number operations.

1. Josie's Jammers have adopted a 4-mile stretch of highway to keep clean. Each afternoon they pick up trash. If they can clean $\frac{2}{3}$ of a mile per day, how many days will it take them to clean the whole section?

2. Chien has $\frac{2}{3}$ of a gallon of gasoline, and each time he mows the lawn, he uses $\frac{1}{6}$ of a gallon. How many times can he mow the lawn before buying more gasoline?

3. Lyra is building dog houses. She has $7\frac{1}{2}$ pounds of nails. Each dog house requires $1\frac{1}{4}$ pounds of nails. How many dog houses can she make?

4. Rita has $3\frac{3}{4}$ ounces of perfume and wants to sell the perfume in $\frac{3}{8}$-ounce bottles. How many bottles of perfume can she make?

5. Jonah has $3\frac{3}{5}$ pounds of dog food. Each day his dog eats $\frac{2}{5}$ of a pound. How many days' worth of dog food does he have?

EXPLORATION 5.15   Remainders

Recall the wire problem (Investigation 1.4B). In many real-life situations, the remainder is as important as the quotient. The following exploration focuses on making sense of remainders with fractions.

Reconciling an Apparent Contradiction

1. Marvin has 11 yards of cloth to make costumes for a play. Each costume requires $1\frac{1}{2}$ yards of material. How many costumes can he make?

    a. Represent and solve the problem with a diagram.

    b. Solve the problem using the division algorithm.

2. Can you explain the fact that the diagram and the algorithm produce different answers?

    Note: Most students find this to be a rather challenging question. Discuss the question with your partner(s) before reading on.

Different Ways to Express Remainders

3. Many students find it helpful to go back and analyze a simpler problem. Consider the following problem: A doctor has 31 ounces of medicine, and each dose is 4 ounces. How many doses can the doctor make?

    a. When doing the computation, we can represent the answer as $7\frac{3}{4}$ or as 7 R3.

    b. What does the $\frac{3}{4}$ mean?

    c. What does the R3 mean?

4. Now go back to Step 2. Write your present thinking about why we get two different answers.

5. When we divide two fractions in which the quotient is not a whole number, sometimes we are interested in the fraction that the algorithm produces (for example, $\frac{1}{3}$ in the Marvin problem). In other cases, we are interested in the fraction that the diagram produces (for example, $\frac{1}{2}$ in the Marvin problem). Make up two story problems, one in which the fraction that the algorithm produces is relevant, and one in which the fraction that the diagram produces is relevant.

    Hint for Brahmagupta's algorithm: In order to answer each question, you need to subdivide the diagram further so that you can repeatedly subtract the second fraction (divisor).

*Mathematics
for Elementary
School Teachers*
p. 279

EXPLORATION 5.16 Meanings of Operations with Fractions[7]

The purpose of this exploration is for you to develop your operation sense by connecting the problem with the operation.

Matching problem situations with operations

For each of the six problems below, do the following:

a. Represent the problem with a diagram.

b. Select the model that fits and briefly justify your choice:

+   combine, increase

−   take-away, comparison, missing addend

×   repeated addition, area, Cartesian product

÷   partitioning, repeated subtraction, missing factor

c. Write a number sentence that would answer the question, *but do not determine the answer.*

For example: A patient requires $\frac{3}{4}$ of an ounce of medicine each day. If the bottle contains 12 ounces, how many days' supply does the patient have?

a.

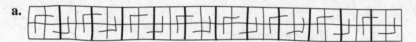

b. I realized I could look at the diagram and ask how many $\frac{3}{4}$s I could take away until I ran out of medicine. Then I realized this was repeated subtraction.

c. $12 \div \dfrac{3}{4}$

Problems

1. The label on a bottle of juice says that $\frac{3}{4}$ of the bottle consists of apple juice, $\frac{1}{6}$ of the bottle consists of cherry juice, and the rest is water. What fraction of the bottle is juice?

2. Freida had 12 inches of wire and cut pieces that were each $\frac{3}{4}$ of an inch long. How many pieces does she have now?

3. Jake had 12 cookies and ate $\frac{3}{4}$ of them. How many cookies did he eat?

4. Kareem had 12 gallons of ice cream in the freezer for his party. Last night Brad and Mary ate $\frac{3}{4}$ of a gallon. How much ice cream is left?

5. The Bassarear family is driving from home to a friend's house, and Emily and Josh are restless. They ask, "How far do we have to go?" Their father replies that they have gone 12 miles and that they are $\frac{3}{4}$ of the way there. What is the distance from home to the friend's house?

6. Karla has $\frac{3}{4}$ of an acre of land for her garden. She has divided this garden into 12 equal regions. What is the size of each region?

EXPLORATION 5.17 Developing Operation Sense

Each part of this exploration is designed to develop both your number sense and operation sense with fractions.

PART 1: Mental arithmetic

Perform each of the computations mentally. Briefly describe your thinking.

1. $2\frac{1}{2} + \frac{3}{4}$

2. $3\frac{1}{4} + \frac{1}{2} + \frac{3}{8} + 2\frac{3}{4}$

3. $8 - 2\frac{3}{5}$

4. $3\frac{1}{8} \times 24$

5. $\frac{2}{3} \times 15$

6. $3 \div \frac{1}{5}$.

PART 2. Using estimation

Determine whether the computation will produce an answer more than 1 or less than 1. Briefly justify your choice.

1. $\frac{11}{12} + \frac{1}{8}$

2. $\frac{7}{8} + \frac{1}{12}$

3. $1\frac{1}{2} - \frac{5}{6}$

4. $3 - 1\frac{3}{4}$

5. $\frac{11}{13} \times \frac{13}{19}$

6. $\frac{7}{8} \times \frac{5}{6}$

7. $\frac{1}{2} \times 1\frac{9}{10}$

8. $\frac{5}{8} \times 2\frac{1}{3}$

9. $\frac{4}{5} \div \frac{2}{3}$

10. $\frac{2}{5} \div \frac{7}{8}$

PART 3: Using estimation and fraction sense

In each case, determine which of the choices represents the better estimate. Explain your reasoning.

1. $3\frac{3}{8} + 6\frac{2}{5} + 9\frac{1}{3}$ Less than 20 or greater than 20?

2. $15\frac{1}{3} - 6\frac{3}{4}$ Less than 9 or greater than 9?

3. $3\frac{3}{4} \times 2\frac{1}{2}$ Less than 8 or greater than 8?

4. $5\dfrac{3}{8} \div \dfrac{1}{2}$ Less than 10 or greater than 10?

5. $3\dfrac{3}{4} \div \dfrac{2}{3}$ Less than 4 or greater than 4?

PART 4: Where is the answer on the number line?

Determine the region in which each answer will lie. Explain your reasoning.

1. $\dfrac{1}{2} \times \dfrac{3}{4}$

2. $\dfrac{1}{2} \div \dfrac{3}{4}$

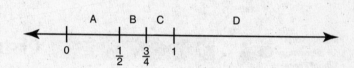

3. $\dfrac{3}{4} \div \dfrac{1}{2}$

4. $\dfrac{7}{13} + \dfrac{12}{17}$

5. $\dfrac{7}{13} \times \dfrac{12}{17}$

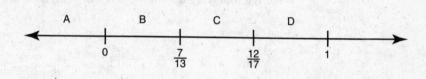

6. $\dfrac{12}{17} \div \dfrac{7}{13}$

PART 5: Where do the digits go?

1. Using four different digits, make two proper fractions whose sum is as close as you can get to 1 but still less than 1.

2. Using four different digits, make two proper fractions whose difference is as great as possible but still a positive number.

3. Using four different digits, make two proper fractions whose product is as close as you can get to 1.

4. Using four different digits, make the smallest possible quotient.

$$\frac{\square}{\square} + \frac{\square}{\square}$$

$$\frac{\square}{\square} - \frac{\square}{\square}$$

$$\frac{\square}{\square} \times \frac{\square}{\square}$$

$$\frac{\square}{\square} \div \frac{\square}{\square}$$

SECTION 5.4   Exploring Beyond Integers and Fractions: Decimals, Exponents, and Real Numbers

There is much evidence that many students' understanding of decimals is mostly procedural. That is, they can compute with decimals, but their understanding of decimals is only weakly connected to their understanding of whole numbers and fractions and weakly connected to what the four fundamental operations mean. The following explorations will help you build stronger connections among many of the key concepts underlying operations with decimals.

Mathematics for Elementary School Teachers
p. 290

EXPLORATION 5.18   Decimals and Base Ten Blocks

Connecting Decimals to Base Ten Blocks

For the following questions, you will need base ten blocks.

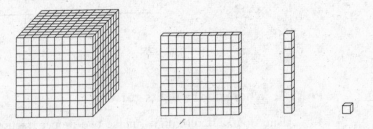

1. If the flat has a value of 1, represent 0.24 in two different ways. Justify your answers.

2. Represent 4.32 in two different ways. Specify your choice of unit for each representation.

3. Represent 0.0127 with base ten blocks. Justify your answer. Specify your choice of unit.

4. Demonstrate 0.463 + 0.507 with base ten blocks. Justify each step.

5. Demonstrate 0.802 − 0.568 with base ten blocks. Justify each step.

6. Show why $0.1 \times 0.1 = 0.01$ with base ten blocks.

7. Demonstrate the equality of 0.1, 0.10, and 0.100 with base ten blocks.

8. Insert $<$ or $>$ between each pair of numbers, and justify your choice:

   a. 0.5   0.28

   b. 8.3   8.14

   c. 2.43   2.4168

   d. 0.07   0.068

   e. 6.74   6.74003

9. Find a decimal between

   a. 1.1 and 1.2

   b. 2.10 and 2.11

   c. 0.99 and 1.0; at least one digit must be other than 0 or 9

10. Write the following decimals in expanded form.

    a. 0.38

    b. 0.026

EXPLORATION 5.19   Exploring Decimal Algorithms

You know *how* to add, subtract, multiply, and divide decimals. This exploration is designed to develop your understanding of and ability to explain *why* those algorithms work.

1. Addition

    a. Below are two examples of common mistakes made by children. What might the children be thinking that would explain why they do it this way?

Problem	Wrong answer	Problem	Wrong answer
$34 + 4.2$	$\begin{array}{r} 34 \\ +\ 4.2 \\ \hline 7.6 \end{array}$	$20.45 + 7.6$	$\begin{array}{r} 20.45 \\ +\ 7.6 \\ \hline 9.645 \end{array}$

    b. In your own words, explain why we need to line up the decimal points when adding decimals.

    c. Explain how understanding of place value helps to understand the algorithm.

    d. Explain how understanding of fractions helps to understand the algorithm.

2. Multiplication

In Chapter 3, you learned that various algorithms for multiplication automatically put the digits of the partial products into the right place. So, too, for multiplication with decimals. We determine the placement of the decimal point by adding the number of decimal places in the multiplicand and the multiplier. Why does this work? That is, why does this algorithm result in all of the digits being in the right place? You can use either or both of the following examples if you wish.

$$\begin{array}{r} 3.6 \\ \times\ 2.4 \\ \hline 144 \\ 72 \\ \hline 8.64 \end{array} \qquad \begin{array}{r} 4.36 \\ \times\ 7.8 \\ \hline 3488 \\ 3052 \\ \hline 34.008 \end{array}$$

3. Division

    a. To divide decimals, you know to move the decimal points in the divisor and the dividend so that the divisor is a whole number. The placement of the decimal point in the quotient is directly above the decimal point in the dividend. Explain why this algorithm works.

$$7.4\overline{)247.9} \qquad \begin{array}{r} 33.5 \\ 74\overline{)2479.0} \\ \underline{222} \\ 259 \\ \underline{222} \\ 370 \\ \underline{370} \end{array}$$

    b. Some children can perform this algorithm when the divisor has fewer decimal places than the dividend, but then they flounder when the divisor has more decimal places than the dividend. Below are two such examples. Why can we just add zeros to the dividend?

$$6.2\overline{)245} \quad \rightarrow \quad 6.2\overline{)245.0} \quad \rightarrow \quad 62\overline{)2450}$$

$$7.54\overline{)46.3} \quad \rightarrow \quad 7.54\overline{)46.30} \quad \rightarrow \quad 754\overline{)4630}$$

EXPLORATION 5.20   Patterns in Repeating Decimals

When we convert some fractions to decimals, sometimes the decimal terminates and sometimes it repeats. For example, $\frac{1}{16} = .0625$, but $\frac{1}{11} = .0909090909090909\ldots$. Rather than writing the "dot dot dot," we place a bar over the portion that repeats, and we present $\frac{1}{11}$ as $0.\overline{09}$.

Note: You might choose to use a spreadsheet to assist in the computations.

PART 1: Predicting the next number in a sequence

1. **a.** The decimal equivalent of $\frac{1}{11}$ is $0.090909\ldots$. Predict the decimal equivalent of $\frac{2}{11}$ and explain your reasoning. Then calculate $\frac{2}{11}$. If your prediction was correct, great. If not, examine the answer so that you can predict the value of $\frac{3}{11}$.

    b. Predict the value of $\frac{3}{11}$ and explain your reasoning. Determine the value of $\frac{3}{11}$. If your prediction was correct, great. If not, determine the value of $\frac{4}{11}$, $\frac{5}{11}$, and so on, until you do see a pattern.

    c. Describe the pattern you see in words.

    d. Predict the value of $\frac{17}{11}$. Explain your reasoning.

2. Determine the decimal equivalent of $\frac{1}{15}$.

    a. Repeat the process described in the first problem:

    Predict and determine the next amount until you can describe the pattern.

    Describe the pattern you see in words.

    b. Predict the value of $\frac{32}{15}$. Explain your reasoning.

3. Determine the decimal equivalent of $\frac{1}{99}$.

    a. Repeat the process described in the first problem:

    Predict and determine the next amount until you can describe the pattern.

    Describe the pattern you see in words.

    Check other fractions in the pattern to see how long the pattern holds.

    b. Predict the value of $\frac{47}{99}$. Explain your reasoning.

    c. Based on your work in the previous problem, predict the value of $\frac{1}{999}$. Explain the reasoning behind your prediction. Then determine the decimal equivalent of $\frac{1}{999}$ and repeat the process used in the previous problems.

    d. Predict the value of $\frac{74}{999}$.

4. Determine the decimal equivalent of $\frac{1}{101}$.

    a. Repeat the process described in the first problem:

    Predict and determine the next amount until you can describe the pattern.

    Describe the pattern you see in words.

    Check other fractions in the pattern to see how long the pattern holds.

    b. Predict the value of $\frac{52}{101}$. Explain your reasoning.

    c. Based on your work in the previous problem, predict the value of $\frac{1}{1001}$. Explain the reasoning behind your prediction. Then determine the decimal equivalent of $\frac{1}{1001}$ and repeat the process used in the previous problems.

    d. Predict the value of $\frac{64}{1001}$.

5. Determine the decimal equivalents of $\frac{1}{7}, \frac{2}{7}, \frac{3}{7}, \frac{4}{7}, \frac{5}{7}, \frac{6}{7}$. Describe the pattern in this sequence of decimals.

6. Predict other families of fractions that might be repeating fractions (e.g., $\frac{1}{11}, \frac{1}{18}$, etc.) Check them out. Describe what kinds of denominators you think might result in repeating fractions, such as odd numbers, and so forth.

PART 2: When will a decimal terminate and when will it repeat?

Now we will investigate a question that fascinated mathematicians several hundred years ago: Can we predict whether the decimal form of a fraction will terminate or repeat?

1. Write down your present thoughts as hypotheses. For example, it will terminate if _____; it will repeat if _____.

2. Determine a plan. For example, you might proceed systematically: $\frac{1}{2}, \frac{1}{3}, \frac{1}{4}$, etc. You might proceed by groups of fractions with common characteristics, such as the denominator is an odd number or prime number, and so on.

3. Collect your data. You might choose to use a spreadsheet to be able to check your hypotheses more quickly. To maximize the development of your mathematical power, use a process like the one shown below. That is, select a fraction, predict whether you think the decimal form will repeat or terminate, and your reasoning. If your prediction was wrong, describe what you learned from analyzing your prediction.

Number	Prediction	Analysis/insight
$\frac{1}{13}$	Repeat because odd number	

4. Present and justify your findings.

Mathematics for Elementary School Teachers
p. 301

EXPLORATION 5.21   Developing Decimal Sense

PART 1: Mental computation

Do each of the following computations mentally. Briefly explain your solution path.

1. $3.65 + 2.4$

2. $0.14 + 0.07$

3. $6 - 1.25$

4. 48×0.5

5. 0.25×60

6. 0.05×280

7. 0.75×36

8. 1.25×16

9. $36 \div .1$

10. $4.8 \div .2$

PART 2: Multiplication sense

These problems are designed to deepen your decimal sense with multiplication. Thus, do the problems without a calculator and by a means other than random guess–check–revise. In each case, briefly summarize your thinking process that led you to a valid answer.

1.   What number times 10 is equal to 62?

2.   What number times 8 is equal to 4.8?

3.   What number times 6 is equal to 0.24?

4.   What number times 0.5 is equal to 4.2?

5.   What number times 0.2 is equal to 4?

6.   What number times 1.5 is equal to 0.45?

7.   What number times 1.5 is equal to 0.036?

8.   What number times .12 is equal to 24?

9.   What number times .15 is equal to 0.045?

PART 3: Division sense

Most students know the rule of moving the decimal point to the left when dividing by powers of 10. That is, $34.2 \div 10 = 3.42$. However, knowing only the rule develops procedural knowledge without conceptual understanding, which, as you have discovered, leads to rented rather than owned knowledge.

We begin here with the simplest cases. We encourage you to think conceptually. For example, what does $42 \div 0.1$ mean? For example, using repeated subtraction, it means how many groups of 0.1 can we take out of 42. For many this is not helpful. Thus, we search for another representation: Applying our knowledge of fractions and the fact that division is the inverse of multiplication, we can see that $42 \div 0.1$ is equivalent to $42 \div \frac{1}{10}$, which is equivalent to 42×10. That is, dividing by 0.1 is equivalent to multiplying by 10.

1. In each case below, the significant digits of the quotients are given. Insert decimals points and 0s when necessary to make each statement true. Briefly summarize your thinking process that led you to the correct answer.

    a.   42   ÷   10   =   42
    b.   42   ÷   0.1   =   42
    c.   42   ÷   0.01   =   42
    d.   3.6   ÷   10   =   36
    e.   3.6   ÷   0.1   =   36
    f.   3.6   ÷   0.01   =   36
    g.   .58   ÷   10   =   58
    h.   .58   ÷   0.1   =   58
    i.   .58   ÷   0.01   =   58

2. In these problems, the significant digits of the quotients are given. Insert the decimal point in the proper place. Briefly summarize your thinking process that led you to the correct answer.

    a.   52.5   ÷   4.2   =   125
    b.   316.8   ÷   66   =   48
    c.   899.6   ÷   26   =   346
    d.   7.82   ÷   3.4   =   23
    e.   89.7   ÷   0.26   =   345
    f.   2.44   ÷   0.08   =   305
    g.   19.47   ÷   0.006   =   3245
    h.   0.448   ÷   0.04   =   112

3. Determine two numbers that will satisfy each of the following problems. Do the problems without a calculator and by a means other than random guess–check–revise. In each case, briefly summarize your thinking process that led you to the correct answer.

    a. $\overset{.5}{0\overline{)0}}$ b. $1.5\overset{0}{\overline{)0}}$ c. $\overset{0}{0\overline{)0.6}}$

4. Determine two numbers that will satisfy each of the following problems; at least one of the numbers must be a decimal. Do the problems without a calculator and by a means other than random guess–check–revise. In each case, briefly summarize your thinking process that led you to the correct answer.

    a. $\overset{.2}{0\overline{)0}}$ b. $0.4\overset{0}{\overline{)0}}$ c. $\overset{0}{0\overline{)0.45}}$

EXPLORATION 5.22   The Right Bucket: A Decimal Game

Materials

- Game sheet, page 131

After playing one or more games, answer the following questions.

1. What did you learn about estimating and mental math with decimals from this game? If you learned specific strategies, describe them.

2. Is it possible to choose the pairs of decimals so that you can get 3 points each time? If you think so, explain why. If you think not, explain why not.

The value of this game comes from refining your ability to estimate. This game comes from *Ideas from the Arithmetic Teacher: Grades 6–8.*[8]

Directions for playing the game

1. Select two decimals from the list given on the game sheet and cross them off the list.

2. Estimate their product and explain your reasoning. An example is shown in the table below.

3. Multiply the two decimals.

4. Determine your score by looking at the bucket chart below. For example, products between 10 and 100 earn a score of 2 points.

5. Select two more decimals, estimate, and then find their product and determine your score. Continue until you have used all the decimals.

6. The goal is to score as many points as you can.

Bucket Chart

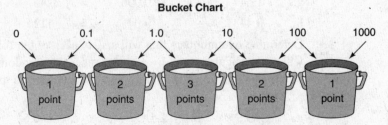

Note: This game can be played competitively (alternating turns) or cooperatively (working together to determine which two decimals make the best pick).

Turn	Decimals	Estimate	Reasoning	Actual product	Points
1	41.2	4.12	0.083 is almost $\frac{1}{10}$, and I knew $\frac{1}{10}$ of 41.2 is 4.12. Thus, I felt that 0.083 of 41.2 would be well over 1.	3.4196	3

Decimals for game 1:

3.6	0.03	13.1	29.6	11.9	0.7	33.7	0.04	21.9	0.125
10.1	0.42	0.07	2.9	0.29	19.5	5.52	23.1	9.6	1.8

Decimals for game 2:

0.023	0.9	5.74	0.245	7.3	1.2	23.8	16.5
0.068	8.7	0.12	1.8	0.78	42.7	0.2	3.6

The Right Bucket Game Sheet for EXPLORATION 5.22

Turn	Decimals	Estimate	Reasoning	Actual product	Points
1					
2					
3					
4					
5					
6					
7					
8					
9					
10					
				Total	

EXPLORATION 5.23   Target: A Decimal Game

Materials

- Game sheet on page 135

Directions for playing the game

Note: These directions are written using an example that involves the operation of multiplication. The game can also be played using any of the other three operations: addition, subtraction, or division.

1. Select a number, an operation, a goal, and a winning zone; for example:

Starting number	145
Operation	multiplication
Goal	1
Winning zone	0.9 to 1.1 (or 0.99 to 1.01, or 1 to 1.1)

2. The first player selects a number to multiply the starting number by, trying to get the product of the two numbers in the winning zone (in this case, between 0.9 and 1.1).

3. If the product is not in the winning zone, then the product becomes the starting number for the second player.

4. Play continues in this manner until the product is within the winning zone.

Directions for playing the game

1. Use the game sheet on page 135. Play the game several times with a partner. Record your game in the table provided on the game sheet. For example, the first turn in the game described in the directions might look like this:

Turn	Computation	Product	Reasoning
1	145×0.006	0.87	I knew that 100×0.01 would be 1. Because 145 is bigger than 100, I picked a number smaller than 0.01.

2. Describe one mental math strategy that you learned during this game.

Target Game Sheet for EXPLORATION 5.23

Starting number   _____

Operation   _____

Goal   _____

Winning zone   _____ to _____

Turn	Computation	Product	Reasoning
1			
2			
3			
4			
5			
6			
7			
8			

EXPLORATION 5.24 A Real-life Problem in College

While in college, I made extra money by typing papers. Let's say you wanted to do this and wanted to make at least $8 an hour. How much should you charge per page? There are two versions of this problem: One is open-ended and one is structured. Choose which version you will solve and then proceed.

1. Open-ended version

    a. Decide what information you need to know in order to solve the problem.

    b. Determine those numbers and justify your process for determining those numbers.

    c. Solve the problem.

    d. Present your solution. Justify each part of your solution.

2. Structured version

    a. Select one person in the class who will determine how many words per minute he or she can type. The class will discuss and agree on the procedure for determining this number so that it will be as accurate as possible. For example, if you determine that this person can type 90 words per minute but when doing actual term papers, that person finds he or she can average only 60 words per minute, the actual hourly rate will be only $\frac{2}{3}$ of what was expected.

    b. Determine how many words, on average, there are in one page of a term paper. As before, determine the procedure for determining this number so that it will be as accurate as possible.

    c. After you have determined the necessary information, solve the problem.

    d. An alternative solution path is to determine, on average, how long it takes for this person to type one page. If your instructor chooses, determine this number also and then compare your answer to the answer you get from using the numbers determined in (a) and (b).

    e. Present your solution.

CHAPTER **6**

Proportional Reasoning

The following explorations require you to grapple with ratios, rates, proportions, and percents, all of which fall under the "big idea" of proportional reasoning. It is ironic that although we probably find the need to use proportional reasoning in our everyday lives as much as we do any other kind of mathematical reasoning, a substantial body of research shows that few adults can competently and confidently navigate nonroutine problems involving proportional reasoning.

SECTION 6.1   Exploring Ratio and Proportion

In this section, you will explore real-life applications of ratio and proportions, examine the connection between proportions and functions, and grapple with the meaning of ratios and proportions.

Mathematics
for Elementary
School Teachers
pp. 316, 324

EXPLORATION 6.1   Which Ramp Is Steeper?

One of the uses of mathematics is to compare. In some cases, the comparison is easy—such as finding who is taller. In the following case, the comparison is not quite so simple. An exploration commonly done with elementary school students has them roll carts down ramps to examine the relationship between the steepness of the ramp and the distance the cart travels. In this exploration, we will focus on one aspect of this experiment—comparing the steepness of different ramps.

Suppose the students weren't told to standardize measurements, and each pair's ramp had different dimensions. For example, one pair had a ramp with a height of 14 cm, a base of 23 cm, and a length of 26.9 cm. Another pair had a ramp with a height of 12 cm, a base of 18 cm, and a length of 21.6 cm. One student hypothesized that the steeper the ramp, the farther the marbles would go, and this prompted a desire to rank the slopes according to steepness. The students didn't want to build the ramps all over again, but each pair had the following measurements: length of base (*B*), height of base (*H*), and length of ramp (*L*).

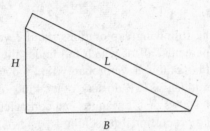

1. Develop a method for determining the relative steepness of the various ramps. *Note:* You are not given the students' actual data. Your job is to develop a plan. This task requires careful thinking, and the following process is suggested:

    a. First, brainstorm different ideas about ways to determine relative steepness of ramps.

    b. Second, discuss the pros and cons of those ideas.

    c. Last, select one idea and discuss a method for determining its validity. That is, what do you need to do in order to determine whether this method is valid?

2. Compare methods that different students have developed. Debate different methods until you come to a conclusion—that is, until you agree on the validity of the different methods. You might want to use the following numbers from two ramps in your debate:

Ramp	Height	Base	Length
A	8 cm	20 cm	21.5 cm
B	12 cm	26 cm	28.6 cm

EXPLORATION 6.2 Using Qualitative Reasoning to Develop Proportional Reasoning

This exploration is adapted from one developed by Esther Billings at Grand Valley State University. My wife and I enjoy eating popcorn when we watch movies at home, and we mix orange juice with seltzer water. I prefer to have more orange juice than my wife.

In each case, juice or water will be added and there are four possible answers:

i. Tom's glass will have a stronger orange juice taste.

ii. Yvette's glass will have a stronger orange juice taste.

iii. The two glasses will have the same taste.

iv. There isn't enough information to know for sure (i.e., we would need to know the actual amounts in each glass).

Situation 1: Tom's glass has a stronger orange juice taste than Yvette's. If we add $\frac{1}{4}$ cup of orange juice to Tom's glass and $\frac{1}{4}$ cup of water to Yvette's glass, which glass will have the stronger orange juice taste now? Explain your reasoning.

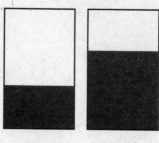

Tom's   Yvette's

Situation 2: Both glasses have the same taste. If we add $\frac{1}{4}$ cup of orange juice to both glasses, which glass will have the stronger juice taste now? Explain your reasoning.

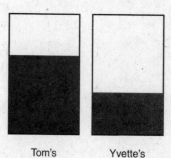

Tom's   Yvette's

Situation 3: Tom's glass has a weaker orange juice taste than Yvette's. If we add $\frac{1}{4}$ cup of orange juice to both glasses, which glass will have the stronger orange juice taste now? Explain your reasoning.

Tom's   Yvette's

Situation 4: Yvette's glass has a stronger orange juice taste than Tom's. If we add $\frac{1}{4}$ cup of orange juice to Tom's glass and $\frac{1}{4}$ cup of water to Yvette's glass, which glass will have the stronger orange juice taste now? Explain your reasoning.

Tom's   Yvette's

'Mathematics
for Elementary
School Teachers
pp. 316, 329

EXPLORATION 6.3   Using Proportional Reasoning to Interpret Data

We are bombarded with numbers and data about our world every day. Sometimes the data are presented in raw numbers and sometimes in rates. The focus of this exploration is to develop understanding of the need for rates, how they are determined, and how they are useful.

PART 1: Deaths from accidents in the United States

1. Take a minute to examine the data in the table below. If someone were to ask you what you see from these data, what would you say?

DEATHS FROM ACCIDENTS IN THE UNITED STATES

Year	Motor vehicles	Falls	Fires, flames, smoke	Drowning	Firearms	Poisoning
1970	54,633	16,926	6,718	6,391	2,406	2,505
2006	44,700	21,200	2,800	3,800	880	25,300

Source: World Almanac, 2009, p. 216.

2. Describe at least two questions you might ask of the person who collected these data.

3. The figures below show the same data but in a slightly different form. Show how the two motor vehicle numbers were determined. The U.S. population in 1970 was 203 million and 300 million in 2006.

**DEATH RATES FROM ACCIDENTS IN THE UNITED STATES
(RATE PER 100,000 POPULATION)**

Year	Motor vehicles	Falls	Fires, flames, smoke	Drowning	Firearms	Poisoning
1970	26.9	8.3	3.3	3.1	1.2	1.2
2006	14.9	7.1	0.9	1.3	0.3	8.4

4. Explain, as if to someone who doesn't understand, what the numbers in the second table mean. For example, what does a motor vehicle death rate of 14.9 mean?

5. Summarize the advantages and disadvantages of using raw numbers or rates to describe a set of numbers.

6. Cecilia has a question. The number of deaths from falls increased, yet the rate declined, from 8.3 to 7.1. So how can the rate have gone down while the actual numbers increased?

PART 2: AIDS in the United States

1. The data below show the estimated number of people living with AIDS in 2006 by ethnicity. Use the numbers from the third column to determine the AIDS rate per 100,000 population.

2. Describe how the rates give a different picture than the raw numbers alone.

Ethnicity	Number of people	Population (millions)
White, not Hispanic	154,495	241.2
Black, not Hispanic	191,590	38.8
Hispanic	80,815	45.5
Asian/Pacific Islander	4,526	13.4
American Indian/Alaska Native	1,651	2.9

Source: Statistical Abstracts of the U.S., 2009, p. 120, p. 11.

PART 3: People in prison

A recent news article reported that the United States had the highest incarceration rate in the world. The first column shows the number of people in prison for nine selected countries. Determine the incarceration rate per 100,000 population and then describe how the rates give a different picture than the raw numbers alone.

Country	Number in 2008	Population (millions)
United States	2,293,157	304
China	1,565,771	1330
Russia	891,738	141
India	373,271	1148
Brazil	440,013	194
South Africa	164,297	49
England	83,392	55
Japan	81,255	127
Egypt	64,378	76

Source: World Prison Population List, 8th Edition, 2008, International Centre for Prison Studies.

PART 4: Greenhouse emissions

By 2006, over 160 countries had ratified the Kyoto Protocol, by which they commit to reduce their emissions of carbon dioxide and several other greenhouse gases or to participate in emissions trading if they are not able to reduce their emissions of these gases by the agreed amount. The United States has been widely criticized for its refusal to ratify the agreement. In the second column below, we see the raw numbers—millions of metric tons of carbon dioxide emitted from fossil fuel consumption in 2005.

In parts 1, 2, and 3, we converted the raw numbers to a rate per 100,000 population. In this case, convert the raw numbers to a rate per 1 million population. Do this and then describe how the rates give a different picture than the raw numbers alone.

Country	Millions of metric tons (2005)	Population (2008) (millions)
United States	5957	304
China	5323	1330
Russia	1696	141
Japan	1230	127
India	1166	1148
Germany	844	82
Canada	631	33

Source: World Almanac, 2009, p. 332.

Mathematics
for Elementary
School Teachers
p. 319

EXPLORATION 6.4   Unit Pricing and Buying Generic

Unit pricing is now common. Underneath most items in grocery stores is the unit price of the item. For example, if a 24-ounce jar of pickles costs $2.79, the unit price is 11.6¢, which means that the pickles cost 11.6¢ per ounce. Most states now require grocery stores to show the unit price below each item. One of the reasons for this law is that people tend to believe that larger items are proportionally cheaper. For example, they believe that a 64-ounce box of detergent will cost less than twice as much as a 32-ounce box. However, many companies use this belief to their advantage.

1. Let's say Paul Bunyan pancake mix comes in two sizes: 30 ounces and 50 ounces. The smaller box costs $2.89, and the larger box costs $4.69.

   a. Without using a calculator, determine which is the better buy.

   b. Use a calculator and record your process. $\frac{2.89}{30} = .7$ $\frac{4.69}{50} = .9$

   c. Jackie did the problem this way: $\frac{30}{2.89} = 10.38$ $\frac{50}{4.69} = 10.66$
   What do 10.38 and 10.66 mean?

   d. What do you think of Jackie's method? Justify your response.

   e. Describe a situation in which you might buy the smaller item even if it cost more proportionally than the larger item.

An issue related to unit pricing is generic products. When I was growing up in the 1950s and 1960s, consumers could choose among the various name brands. Now, however, consumers can choose among the name brands or choose the generic alternative.

1. Are there products for which you are more likely to buy the generic brand or are not likely to buy the generic brand? What reasons do you give for choosing to buy the generic brand or choosing to buy the name brand?

2. One of the obvious reasons for buying the generic brand is that it is cheaper. Let's say you go to a store, and there you see a dispenser of Scotch brand tape that sells for $1.69 and a generic alternative that sells for $1.29. Both rolls contain the same amount of tape.

   a. How could we compare the two prices? Write down your thoughts.

   b. After hearing other strategies, describe and critique different ways in which we could compare the two prices.

3. Let's extend this question.

   a. Select a drugstore and collect and compare data on a name-brand item and a generic brand (or compare newspaper ads). Suppose you were writing an advertisement for the drugstore and you wanted to convince shoppers that they could save a lot of money by buying the generic item. Write the ad.

   b. Gather data on different sizes. Do you always get more for your money with the bigger size?

   c. Gather data from two different stores. How much cheaper is one product at one store than the other?

4. Let's say an average family of four decided to "buy generic" whenever possible. Over the course of a year, how much money would they save?

EXPLORATION 6.5   Proportional Reasoning and Functions

In this exploration, we will explore a variety of functions that rely on proportional reasoning.

1. Solve each of the following problems.

    a. It costs Rita 50¢ for the first 3 minutes of a long-distance call to her boyfriend and 20¢ for each additional minute. If Rita calls her boyfriend and talks for 12 minutes, how much does the call cost?

    b. At Sam's Submarine Sandwich Shop, the cost of your submarine sandwich is determined by the length of the sandwich. If Sam charges $2 per foot, how much will a 20-inch sandwich cost?

    c. The first-grade class is measuring the length of a dinosaur by using students' footsteps. The first dinosaur is 6 of Alisha's footsteps or 9 of Carlo's footsteps. If the second dinosaur is 8 of Alisha's footsteps, how many of Carlo's footsteps will it be?

    d. A farmer has determined that he has enough hay for 4 cows for 3 weeks. If the farmer suddenly obtains 2 more cows, how long can he expect the hay to last?

    e. Certain bacteria can double in number in 1 hour. If we start with 1 bacterium, how many bacteria will there be after 20 hours?

    f. Consider the set of quadrinumbers below. How many dots does the fifth quadrinumber contain?

2. Discuss with your partner(s) solutions and strategies for each of the six problems in Step 1.

3. Determine a general formula for each problem below. Then discuss solutions and strategies for each problem.

    a. If it costs 50 cents for the first 3 minutes and x cents for each additional minute, and a person talks for y minutes, what is the cost?

    b. If the sandwich costs x dollars per foot and it is y inches long, what is the cost?

    c. If the first dinosaur is 6 of Alisha's footsteps or 9 of Carlo's footsteps, and another dinosaur is x of Alisha's footsteps, how many of Carlo's footsteps will it be?

    d. If there is enough hay for 4 cows for 3 weeks, how long would the hay last if there were x cows?

    e. If bacteria double in number every hour, how many bacteria will there be after x hours? Start with 1 bacterium.

    f. How many dots are in the xth quadrinumber?

4. For each problem, construct a graph that shows the relationship between the variables. Describe each graph as though you were talking to someone on the phone.

5. Describe the similarities and differences you see among these six problems. For example, the graphs of some families are straight lines and the graphs of others are not.

SECTION 6.2   Exploring Percents

Percents are a powerful tool that enables us to compare amounts and to describe change.

Mathematics for Elementary School Teachers
pp. 342, 343

EXPLORATION 6.6   Percents

Sales

John sees that the local department store is having a sale. He goes to the store and finds that all televisions are 25% off.

1. Describe in words what that means.

2. Let's say he is interested in a particular television that normally sells for $400. If it is priced at 25% off, how much will he pay for it?

3. There are two common ways in which students solve this problem.

    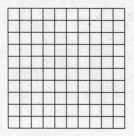

    - Ann: 25% of $400 is $100. $400 − $100 = $300. John pays $300.

    - Bela: 75% of $400 is $300. John pays $300.

    Ann doesn't understand what Bela did. How can you help her? You may, if you wish, use a grid like the one at the right.

4. Let's say that Joe gets a 5% raise and that he presently makes $8.00 per hour. One way to determine his new wage is to find 5% of $8.00 and then add that to $8.00. Using the ideas generated above, can you figure out how to determine his new wage with only one calculation? Describe the method.

Percent Decrease and Increase

Joshua is confused. He works for the Adamson Printing Company. Last year the economy was in such bad shape that all employees agreed to take a 20% cut in pay. However, this year the economy had improved so much that the company agreed to give everyone a 20% raise. Before the pay cut, Joshua was making $30,000 a year.

5. Explain why the 20% raise does not "undo" the 20% pay cut.

6. What raise would undo the 20% pay cut?

7. Determine a general formula that will tell you what percent increase will undo an x percent decrease.

Changes in Rates

In New Hampshire, where I live, there was a tremendous amount of controversy over the Seabrook nuclear power plant. The construction of the plant was held up many times. Eventually the company that made the reactor went bankrupt. It was bought out by another utility company. Part of the deal made to ensure that the new company would make a profit was that it could increase rates by *at least* 5.5% each year for 7 years.

8.   Bill and Betty Olsen figured that their average monthly electric bill last year was $83.21.

   a. If they use, on average, the same amount of electricity over the next 7 years, and their bill increases exactly 5.5% each year, what can they expect their monthly utility bill to be at the end of 7 years?

   b. Meet with your partner(s) and discuss answers and solutions. If you think you would change your method in order to do a similar problem, describe how and why the new method works.

   c. Jarrad used the following method:

$$83.21 \times 0.055 = 4.57655$$
$$83.21 + 4.58 = 87.79$$
$$87.79 \times 0.055 = \text{etc.}$$

Do you think Jarrad's method is valid? Why or why not?

If not, what suggestions would you give to Jarrad? Justify your suggestions.

9.   Last year the Olsens' combined income was $62,310. If their income increases by 5.5% each year, what will their income be at the end of 7 years?

EXPLORATION 6.7   Do You Get What You Pay For?[1]

There are a number of items that you buy by the pound, although you are paying for parts of the items that you don't use. For example, you throw out the shells from peanuts, the husks and cobs from corn, the rind of the watermelon, the bones from chicken, and the peels from oranges. Just what percent of the product are you tossing out?

1.   Let us begin with oranges. In this step, we will only plan a procedure. Data collection takes place in Step 2.

   a. Design a procedure to find out what percent of the weight of an orange is peel.

   b. What assumptions did you make in designing your procedure? (For example, not all oranges have the same percentage of peel; like humans, some have thicker skins!)

   c. Exchange descriptions of the procedure with another person (group). Make comments on the validity of that person or group's procedure and on the clarity of the description. Listen to the comments about your procedure.

   d. What changes would you make in your procedure now that you have compared it with one or more other procedures?

2.   Using the procedure designed in Step 1, carry out the experiment to find the percentage of an orange that you throw away.

   a. Describe your data. Show your results.

   b. Compare your results with those of other groups or of the whole class.

   c. Is there anything that you would change about the design of your experiment? If so, explain your changes.

   d. What did you learn from this exploration?

Extensions

3.   Do you think the grower makes more or less profit from thick-skinned oranges than from thin-skinned ones? Explain your reasoning.

4.   Do you think grapefruits, in general, would have a higher percentage of peel than oranges? What about lemons? Explain your answer.

5.   Describe how you would find the percent of waste for peeled apples.

6.   Determine the cost per pound (paid at the supermarket) and actual cost per pound (when the nonedible portion is discarded) of various items.

   a. Which has the biggest difference in cost per pound?

   b. Which has the biggest percent change?

*Mathematics
for Elementary
School Teachers*
p. 346

EXPLORATION 6.8 Reducing, Enlarging, and Percents

PART 1: A broken copy machine[2]

Most copy machines allow you either to enlarge or to reduce a copy. Some machines let you determine the exact amount of enlargement or reduction. Other machines have buttons for the changes that are most commonly made. Let's say a copy machine has buttons that will enable you to make the following changes to a copy: 10%, 50%, 100%, and 200%. For example, the 10% button means that the size of the copy will be 10% of the size of the original, whereas the 200% button will make a copy that is 200% (double) of the size of the original. The 100% button is what you push when you want a copy that is the same as your original.

1. Let's say the buttons on the machine were as shown below.

   a. How could you make a copy that was 25% of the size of the original?
   b. How could you make a copy that was 3 times the size of the original?
   c. Suppose the 100% button was broken. How could you make a copy of the original that was the same size?

2. Suppose these were the buttons on a machine, and the 100% button was broken. How could you make a copy of the original that was the same size?

3. Suppose these were the buttons on a machine, and the 100% button was broken.

   a. Explain why it is now impossible to make a copy of the original that is the same size.
   b. How close can you get to the original size?

PART 2: Making all the quilt blocks the same size

I encountered this problem while I was making the quilt patterns in Chapters 8, 9, and 10. I was able to make the patterns using a graphing software program. Most of the quilt patterns that I made are based on a grid. That is, each pattern can be achieved by making a 3 × 3, a 4 × 4, a 5 × 5, or a 6 × 6 grid. To make the quilt blocks, I first made a 6 × 6 grid on my computer. When I went to make a specific block, I opened the file containing the 6 × 6 grid and used whatever I needed. Therefore, all the little squares were always the same size.

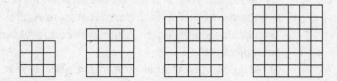

The problem I encountered is that when I wanted to print the blocks, I wanted them to all be the same size and to be smaller than the size that showed on the screen. I determined that I could get the desired size for the 6 × 6 block by telling the computer to reduce the quilt block to 25%.

1. What did I need to tell the computer so that the reduced 5 × 5 block would be the same size as the reduced 6 × 6 block?

2. What did I need to tell the computer so that the reduced 4 × 4 block would be the same size as the reduced 6 × 6 block?

3. What did I need to tell the computer so that the reduced 3 × 3 block would be the same size as the reduced 6 × 6 block?

4. I made a block starting with an 8 × 8 grid. What would I need to tell the computer so that the reduced 8 × 8 block would be the same size as the reduced 6 × 6 block?

EXPLORATION 6.9 Mice on Two Islands

There are two different islands in the Southern Micific Ocean, Azumi and Muremi, each of which has 20,000 mice. In this exploration, we will explore different ways in which populations can change.

1. Brainstorm factors that could affect the rate of growth of the mouse populations.

2. Let's say that the factors on Azumi produce an environment in which the population grows by 3000 each year, and the factors on Muremi produce an environment in which the population grows by 10% each year. That is, in Azumi's case, we can determine the next year's population additively; in Muremi's case, we can determine the next year's population multiplicatively.

   A quick computation shows that after 1 year, Azumi will have more mice—23,000, compared with Muremi's mouse population of 22,000. Do you think that Azumi will always have more mice than Muremi (disregard such factors as disease and predators)? Why or why not?

3. **a.** Determine a method for finding the annual mouse populations of the two islands for the next 15 years.

   b. Determine the annual populations and compare answers and solutions with your partner(s).

   c. If you found errors in your method, note them and explain the cause of the error. If you find that you prefer someone else's method, explain and justify the new method.

4. Let us focus now not on the numbers of mice on each island but on the relationship between the two populations. We can then use this information to extrapolate into the future.

   a. Let us examine two different ways to compare the relative sizes of the populations of the two islands. We could make a table showing the difference between the two populations, or we could make a table showing the ratio of the two populations. Make the two tables, and then graph the results on two different graphs. Explain each graph as though you were talking to someone who missed this exploration.

   b. Each graph is useful for answering different kinds of questions. For example, which graph would you use to answer the following question: How much larger will Muremi's mouse population be in 5 years? Which graph would help you predict when Muremi's mouse population will be 50% greater than Azumi's mouse population? Explain your answers.

   c. Compare your predictions with those of your partner(s). As always, if you want to change something as a result of the discussion, identify the change you wish to make and justify it.

CHAPTER 7

Uncertainty: Data and Chance

The explorations in this chapter have been designed to provide you with meaningful experiences involving the collection and interpretation of data and probability experiments so that you can grapple with some of the basic ideas in the fields of statistics and probability. I want you to experience the excitement that I have seen elementary and college students feel when they suddenly see something from graphing a set of data that they didn't see simply from looking at the numbers, or when they see patterns in chance phenomena that enable them to have a better sense of the probability of that event happening. Note that there were explorations in Chapter 2 where you collected data and that there are explorations in Chapter 10 where you will collect data.

SECTION 7.1   The Process of Collecting and Analyzing Data

In order to understand the big ideas of statistics and to develop competence in collecting and analyzing data, it is helpful to go through the process of defining a question, collecting and analyzing the data, and then presenting your conclusions. It is also helpful to be able to critically analyze data that other people have collected. In these explorations, you will grapple with the important ideas of statistics: defining the question carefully, understanding variation and how to minimize measurement variation, the shapes that data can take, different ways to graph data, and how to represent and interpret centers and the spread of a set of data.

*Mathematics
for Elementary
School Teachers*
pp. 358, 379

EXPLORATION 7.1   Population Growth and Density

Did you know that the framers of the U.S. Constitution required, in Article 1, Section 2, that a census be taken every 10 years?

Population Growth

1. a. Look at the data at the right and write down your initial impressions.

    b. Predict the population in the year 2010 on the basis of these numbers. Briefly explain how you came up with your number.

2. Now let us explore how a graph helps us to see a set of data in a different light.

    a. Select and make a graph for the population data from 1790 to 2000. Describe any problems you had in constructing the graph and how you solved these problems—for example, choosing a scale that would enable you to fit numbers as low as 3 million and as high as 281 million.

    b. Does the graph help you to see the population growth in a different light? If so, briefly explain.

    c. Now predict the population in the year 2010 on the basis of your impressions of the graph. Briefly describe how you obtained the number. Compare this number to your prediction in Step 1(b).

3. Now let us examine how graphs can enrich our understanding of the growth in population.

    a. Make two line graphs, the first graph showing the actual increase in population from decade to decade and the second graph showing the percent increase in population from decade to decade.

    b. The first of these graphs shows population growth from an additive perspective, the second from a multiplicative perspective. Describe the different impressions that these graphs give. Summarize the advantages and disadvantages of displaying the population increase in each of these ways.

    c. Describe the growth of the U.S. population over the last 50 years. Then compare your response to the one you gave in Step 1(a).

Population Density

4. a. Focus now on the "Population per square mile" column. Describe your first impressions, and note any questions you have about the data.

    b. Suppose someone came up to you and said, "I don't see the point of these numbers. What do they tell you, anyway?" How would you respond to that person?

    c. How do you think the Census Bureau came up with the number 79.6 for the year 2000? In other words, what mathematics do you think it used?

U.S. POPULATION, POPULATION DENSITY, AND AREA OF RESIDENCE, 1790–2000

Year	Total population	Percent increase	Pop. per sq. mi.
1790	3,929,214	N.A.	4.5
1800	5,308,483	35.1%	6.1
1810	7,239,881	36.4	4.3
1820	9,638,453	33.1	5.5
1830	12,866,020	33.5	7.4
1840	17,069,453	32.7	9.8
1850	23,191,876	35.9	7.9
1860	31,443,321	35.6	10.6
1870	39,818,449	26.6	13.4
1880	50,155,783	26.0	16.9
1890	62,947,714	25.5	21.2
1900	75,994,575	20.7	25.6
1910	91,972,266	21.0	31.0
1920	105,710,620	14.9	35.6
1930	122,775,046	16.1	41.2
1940	131,669,275	7.2	44.2
1950	150,697,361	14.5	50.7
1960	179,323,175	18.5	50.6
1970	203,302,031	13.4	57.4
1980	226,545,805	11.4	64.0
1990	248,709,873	9.8	70.3
2000	281,421,906	13.2	79.6

Source: From *The Universal Almanac.* Copyright © 1992 by John W. Wright. Reprinted with permission.

*Mathematics
for Elementary
School Teachers*
pp. 360, 367

EXPLORATION 7.2   **Collecting Data to Understand a Population: Typical Person**

The word *typical* conjures up many of the same thoughts that the word *average* does. In this exploration, you will develop a profile of the typical person in your class by analyzing data that you and your classmates decide to collect.

1. Briefly describe topics about which you would like to collect data—for example, number of siblings, place of birth, number of schools attended in grades K–12, favorite TV show, and so on.

2. Having selected a topic, come up with the question you will ask your classmates. You want to make sure that the question is clear and that everyone will interpret the question in the same way. For example, if your topic is "number of brothers and sisters," how might others interpret this statement in different ways?

3. Analyze your data.

   a. First, discuss what kind of graph you want to make. Make the graph and briefly describe why you chose this graph over other graphs.

   b. Determine the mean, median, and mode, as appropriate. How did you determine them? Does one seem to be a more appropriate measure of the typical student than the others? Explain.

   c. Identify any challenges you had with the question, with the data, or with making the graph. Describe what each challenge was and how you met the challenge.

4. Prepare a presentation for the class. Include your question, your results from Step 3, and the words you will say in your presentation.

Looking Back

1. *Looking back on your presentation*

   a. Describe the strengths of your group presentation.

   b. Decide whether any of the following changes would make your presentation better: improvements in your graph, a different graph, changes in your presentation of your graph, or your determination and/or interpretation of mean, median, and mode.

2. *Looking back at other presentations*

   a. Select the group presentation that you thought was the best. Describe what was so good about that presentation.

   b. Select a presentation in which the students chose a graph that you did not think was the best choice for those data. Describe the topic and graph. Indicate what kind of graph you think would be more appropriate, and explain why.

*Mathematics
for Elementary
School Teachers
pp. 368, 375*

EXPLORATION 7.3   Exploring the Concept of Average

The notation of median and mode are intuitively constructed by children. However, the mean is neither intuitively constructed nor understood beyond the procedure used to obtain it. Developing a deeper understanding of the mean is the goal of this exploration.

PART 1: What does the mean mean?

Imagine a class of 6 students being asked how many movies they saw last month, and we know that the mean was 4.0. The scenario at the right is possible, but unlikely.

```
                        x
                        x
                        x
                        x
                        x
                        x
0  1  2  3  4  5  6  7  8  9  10
```

1. What if we know that the mean is 4 and we know five of the numbers. What would the sixth number be? Explain your response without computing the mean.

```
                        x
                        x
                        x
               x        x
0  1  2  3  4  5  6  7  8  9  10
```

2. What if we know that the mean is 4 and we know four of the numbers. What could the other numbers be, other than 6 and 6? Why?

```
         x        x
         x        x
0  1  2  3  4  5  6  7  8  9  10
```

3. What if we know that the mean is 4 and we know four of the numbers. What could the other two numbers be? Why?

```
               x
               x
               x                    x
0  1  2  3  4  5  6  7  8  9  10
```

4. **a.** Make a scenario that would have a mean of 4 in which none of the data are 4.

```
0  1  2  3  4  5  6  7  8  9  10
```

   b. Such a distribution is not unusual, but it is bothersome for many children, who argue, "How can the average be 4 when no one saw 4 movies?" How would you respond?

5. **a.** Make a scenario that would have a mean of 4 in which 2 people saw no movies.

```
0  1  2  3  4  5  6  7  8  9  10
```

   b. Such a distribution is not unusual, but it is bothersome for many children, who argue that zero doesn't count. How would you respond?

Mean as Balance Point

Construct a balance with a ruler, pencil, tape, and pennies. Tape the pencil to the table, and then tape the ruler to the pencil (double-sided tape works best). Place 6 pennies on the ruler at the 6-inch point, and make sure the ruler balances. If we now move one penny to the 4-inch point, and another penny to the 8-inch point, the ruler still balances.

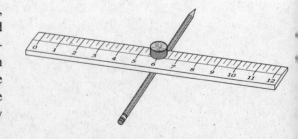

6. Move 4 coins from the 6-inch point to the other points so that the ruler still balances. Write the numbers. Explain, in numerical terms, why it still balances without adding all the numbers and dividing by 6.

7. Make another arrangement that balances but is not symmetric and explain why it still balances.

8. Without adding the numbers and dividing by the sum of the numbers, predict whether the ruler will balance for the pennies located at the following spots. Explain your reasoning. If it doesn't balance, change the location of one penny to make it balance.

    a. 4   4   7   9

    b. 5   5   5   9

    c. 2   2   2   11

Mean as Fair Share

9. Let's say we determine how many pencils each child has in his or her desk and we get the results shown at the right. That is, one child had 2 pencils, another 3, another 4, and so on. Give pencils from those who have more to those who have fewer until everyone has the same number of pencils. How many pencils does each child have now?

10. Turn to the second arrangement. Again redistribute the pencils so that every child has the same number.

11. What does redistribution have to do with mean?

```
                                    x
                                   x x
                                  x x x x
                                 x x x x x
                                x x x x x x
                                x x x x x x

                                    x
                                   x x
                                  x x x
                                  x x x
                                  x x x
                                 x x x x
                                 x x x x
```

Putting It Together

12. Imagine the exam scores of a class of 10 students. Make three very different distributions that all have a mean of 80.

PART 2: Mean, median, and mode

A class of 11 college students has been surveyed and asked how many drinks they had in the past week. Below are several sets of data, each for a group of 11 students.

1. In each case, determine the mean, median, and mode:

    a. 0, 0, 2, 3, 5, 6, 7, 15, 17, 20, 35

    b. 0, 0, 0, 0, 0, 0, 15, 20, 20, 25, 30

    c. 3, 5, 5, 5, 7, 9, 12, 13, 15, 15, 17

    d. 0, 3, 4, 4, 10, 10, 10, 15, 15, 17, 22

    e. What do you conclude, from these four sets of data, about what the mean, median, and mode tell you?

2. **a.** Create 10 numbers where the mean and median are both greater than 10 and the mode is less than 10.

    b. Create 10 numbers where the mean and mode are both greater than 10 and the median is less than 10.

    c. Create 10 numbers where the mode and median are both greater than 10 and the mean is less than 10.

3. Explain what you learned from questions 1 and 2.

4. The instructor has graded the exam and tells you that the mean is 78. The instructor has all the data and has done every analysis you can think of.

   a. Write down what comes to your mind when you hear only this number.

   b. What does this number not tell you?

   c. You can ask for one additional piece of information. Your goal is to understand better how the class did overall. What additional information would you like—why?

   d. You can see one graph. What graph would you like? Why?

5. You have applied for a job in a far-away state, and lots of information is available on the Web. You find that the average salary for a new teacher in that state is $46,250.

   a. Write down what comes to your mind when you hear only this number.

   b. What does this number not tell you?

   c. If you could dig up one additional piece of information about salaries in that state, what would it be? Why?

   d. Your instructor has given you that information. What did it tell you? What did it not tell you?

   e. If you could dig up one additional piece of information, what would it be? Why?

   f. Your instructor has given you that information. What did it tell you? What did it not tell you?

6. You are going to write a newspaper story on the exercise habits of students at a high school whose enrollment is 842. The statistics have to be kept short. You have room for one graph and 2 or 3 sentences. What graph would you want and what information would you include in those 2 or 3 sentences? Justify your choice.

7. Describe the understanding of *mean* that you have gained from this exploration. Consider, for example, what it means, what it tells us, what it doesn't tell us, and the like.

Mathematics or Elementary School Teachers pp. 359, 369

EXPLORATION 7.4   Explorations for Gathering and Analyzing Data

In most of the cases in the exploration, you will be asked to describe how you minimized measurement variation. Therefore, take some time before you record your official data to make sure that your procedure is as reliable (consistent) as possible.

NUMBER 1: How long is your reaction time?

Materials

- Rulers

Measuring reaction time is important for driving safety and for many occupations, for example, jet pilots. While there are many high-tech ways to measure one's reaction time, here is a low-tech way.

Procedure: Students work in pairs. The first person holds a ruler upright. The second person positions his/her fingers on either side of the ruler a specified distance apart. Without warning the first person lets go of the ruler and the second person tries to grab the ruler as quickly as possible. By subtracting the number where the person grabbed the ruler from the original number, we get a measure of the person's reaction time.

1.   Describe what you did to minimize measurement variation.

2.   Gather data for each person until you are confident that the number you have is the "true" reaction time for that individual. Justify your decision about how you determined that number.

3.   Analyze the data for the whole class. What is the average reaction time? What else can you conclude from the data?

4.   What did you learn?

NUMBER 2: How does more materials make the bridge stronger?

Materials

- Spaghetti, Styrofoam cups, pennies

Poke two holes on either side of the cup. Insert one strand of spaghetti through the holes and suspend the cup, either by one person gently holding the spaghetti on either side of the cup or by having two stacks of books and then suspending the spaghetti bridge between the two stacks.

1.   Describe what you did to minimize measurement variation.

2.   Gently drop pennies into the cup until the bridge breaks. Repeat the process until you are confident that you have minimized the measurement variation.

3.   Repeat the process with a bridge of two strands, three strands, etc.

4.   Analyze your results.

5.   What did you learn?

NUMBER 3: How does weight affect the length of the jump?

Materials

- Rubber bands, paperclips, weights (20 heavy washers work nicely), ruler, plastic bags

Assemble your bungee jump.

1. Describe what you did to minimize measurement variation.
2. Place one weight in the bag and record how much the bungee cord increased in length.
3. Repeat this five times, adding one weight each time.
4. Predict the result for 10 weights and describe your reasoning.
5. Gather more data.
6. Compare your result to your predicton.
7. Predict the result for 20 weights and describe your reasoning.
8. Using the same equipment, repeat the experiment one or two days later!
9. What did you learn?

NUMBER 4: How many times can you snap your fingers?

Materials

- A watch or timer that can record 30 seconds

1. Discuss what counts as a "snap."
2. Have each person snap his or her fingers as quickly as possible for 30 seconds.
3. Analyze the data.
4. Let people practice.
5. Repeat the process.
6. Compare this set of data with the first set of data.
7. What did you learn?

NUMBER 5: How good are you at penny horseshoes?

Materials

- Pennies and a noncarpeted floor

Stand a designated distance from the wall and toss a penny toward the wall. The goal is to have the penny land as close as possible to the wall.

1. Describe what you did to minimize measurement variation.
2. Have each person do this a determined number of times.
3. Collect all the data.
4. Analyze the results.
5. What did you learn?

SECTION 7.2   Going Beyond the Basics

EXPLORATION 7.5   How Many Drops of Water Will a Penny Hold?

Materials

- Pennies, eyedroppers, paper towels

  Gently drop one drop of water at a time onto a penny until the water spills.

1. Spend some time thinking about how to minimize measurement variation. That is, ideally, you should be able to get the same number of drops each time. Describe what you did to minimize measurement variation.

2. Describe your procedure precisely, so someone reading your description could replicate what you did.

3. Do the experiment at least 20 times. Did the variation of the data decrease as you gathered more data? That is, did you become better at the process?

4. Analyze and present your results. Include:

   a. what you learned about minimizing measurement variation

   b. the analyses you did of your data: computations and graphs

   c. your conclusions

5. Repeat this experiment with quarters. First, based on your penny data, predict how many drops a quarter will hold.

6. Compare the penny and quarter data. How are they similar and how are they different?

EXPLORATION 7.6   **How Accurate Can You Get the Whirlybird to Be?**

Materials

- Whirlybirds and measuring sticks or measuring tape. Make your own whirlybird. You can get the diagram from this website: http://www.pbs.org/teachers/mathline/lessonplans/esmp/whirlybird/whirlybird_procedure.shtm

Hold the whirlybird about three or four feet above the ground.
Mark the point directly below your hand.
Drop the whirlybird and record how far it fell from that point.

1. Spend some time thinking about how to minimize measurement variation. Describe what you did to minimize measurement variation.

2. Describe your procedure precisely, so someone reading your description could replicate what you did.

3. **a.** Drop the whirlybird 15 times and collect the data.

   b. Discuss how you can improve the accuracy of the drops.

   c. Drop the whirlybird 15 more times and collect the data.

4. Analyze and present your results. Include

   a. What you learned about minimizing measurement variation

   b. The analyses you did of your data: computations and graphs

   c. How the first and second sets of data are alike and how they are different

   d. Your conclusions

EXPLORATION 7.7   **Exploring Relationships Among Body Ratios**

Materials

- Measuring tape, string, rulers

Having students gather and analyze data about various measurements of their bodies is an exploration that has appeared in many publications. This exploration has wonderful real-world applications, too. Forensic examiners and palentologists can give a very good approximation of the height of a person just from knowing the length of certain bones.

1. Using measuring tape, measuring sticks, and string, gather as much of the following data from each person in centimeters as your instructor directs.

A	Height			
B	Floor to belly button			
C	Belly button to top of head (A − B)			
D	Arm span			
E	Head circumference			
F	Radius			
G	Humerus			
H	Femur			
I	Tibia			
J	Foot			
K	Wrist			
L	Thumb			
M	Neck			
N	?			

Radius: Measure from the elbow bone to the wrist bone.

Humerus: Measure from the elbow bone to the top of the shoulder.

Femur: Measure from the top of the hip bone to the middle of the knee cap.

Tibia: Measure from the middle of the kneecap to the ankle bone.

Palm span: Holding the palm gently closed, measure the distance across the palm.

Hand span: Holding the fingers as far apart as possible, measure the distance from the tip of the little finger to the tip of the thumb.

2. **a.** Select one set of measurements (e.g., height). Just from looking at the data, what do you see? What do you think the line plot or histogram for that data will look like?

    b. Analyze the data. Summarize what you found about that population: measures of central tendency, shape of the data, variation, etc.

3. **a.** Determine those ratios below selected by your instructor. You can use a spreadsheet to make this task less tedious!

A:B   A:D   A:E   A:F   A:G   A:H   A:I   A:J

C:B   K:L   M:K

b. Analyze your data.

c. Which ratios show the least variation?

4. Use the following formulas to determine each person's predicted height from your data.

a. Determine the percent error in each case.

b. Which bone seems to work best?

Male Height	Female Height
$2.9 \times$ length of humerus $+ 70.6$	$2.8 \times$ length of humerus $+ 71.5$
$3.3 \times$ length of radius $+ 86.0$	$3.3 \times$ length of radius $+ 81.2$
$1.9 \times$ length of femur $+ 81.3$	$1.9 \times$ length of femur $+ 72.8$
$2.4 \times$ length of tibia $+ 78.7$	$2.4 \times$ length of tibia $+ 74.8$

5. **a.** Make a scatterplot for as many of the pairs of variables as your instructor directs. You can use a spreadsheet to make this task less tedious!

b. Which scatterplots show the highest positive correlation?

6. In *Gulliver's Travels,* the Lilliputians had to make clothes for Gulliver. In the book, it is said that the Lilliputians only needed to measure Gulliver's thumb circumference in order to get his wrist, neck, and waist circumference, because each measure was twice the preceding measure. How true does that generalization hold for this class?

7. A crime has been committed, and forensics people have recovered a handprint from the suspected culprit. If the length of the middle finger on the handprint is 8.9 centimeters, and they are pretty sure the suspect is an adult female, how tall do you think the suspect is?

8. Archaeologists have unearthed a statue of an adult female from ancient Greece. Unfortunately, both of the arms are missing. If the statue is 8 feet 6 inches tall, how long should the arms be?

9. Jack climbed the beanstalk and fell into a footprint of a giant. If the giant's footprint is 44 inches long, how tall is the giant?

10. A forensics team has unearthed a partial skeleton of an adult female from ancient Egypt. The length of the tibia bone (which goes from the middle of the kneecap to the ankle bone) is 36.5 centimeters. How tall was the person?

EXPLORATION 7.8   **Collecting Data to Make a Decision: What Container Best Keeps Coffee Hot?**

The question: What container(s) will keep the coffee warmer the longest?

Materials

- Different kinds of coffee cups—styrofoam, ceramic, metal, and so on; thermometer or temperature probe; and clock or watch

  Preferable: microwave so you can ensure that all liquids begin at the same temperature

Before you start

Discuss and decide on your procedure. One approach is to bring thermoses and pour liquid into each cup, thus ensuring that all begin at the same temperature. If you are using a temperature probe, the data collection is continuous. If using thermometers, determine the interval for collection: 30 seconds, 1 minute, or the like.

1. Examine the objects you have.

    a. Predict the order in which the cups will retain the heat.

    b. Predict the shape of the graphs of time vs. temperature. That is, sketch what you think the graphs will look like, and explain why you think the graphs will have that appearance.

2. Collect the data for the various cups.

3. Analyze the data.

    a. Represent the results for each cup graphically.

    b. Do the graphs add to your ability to say which container is best and whether any one container works better than others? Or do the raw data work just as well?

    c. Decide what other analyses you want to do with the data.

4. Identify and discuss the factors that could affect the accuracy of your data.

5. Report your answer to the question.

6. *Extension:* How will a lid slow down the rate of cooling? Take two cups of liquid in two identical containers, only one of which has a lid. Now measure the rate of cooling. Describe the results.

EXPLORATION 7.9 Explorations for Comparing Two Sets
of Data

NUMBER 1: How well can you estimate the length of an object?

Materials

- One or more lengths of string

1. The instructor will hold up a string of a certain length. Each student silently writes down his or her estimate of the length.

2. Students describe the reasoning behind their estimate until all the different methods and ideas are stated.

3. Once again, each student silently writes down his or her estimate of the length.

4. Analyze the two sets of data.

5. Get the actual length. Is the second set of scores better? If so, how much better?

6. Repeat the process with a string of different length.

NUMBER 2: How does exercise affect pulse rate?

Materials

- A timer that can measure seconds

1. Have each person take his or her pulse while sitting down.

2. Have each person do the same amount of exercise for a specified period: walking in place, jumping jacks, and so on. *Note:* Any person with health concerns should abstain from this.

3. Have each person take his or her pulse at the end of this period. Describe and justify your method for determining your pulse.

4. Analyze and present the results.

NUMBER 3: How many peanuts can you hold in one hand?

Materials

- Peanuts and bags or containers to hold the peanuts

Determine what two groups you will have: students in two classes; students in your class and a group of elementary school children, etc.

1. Spend some time thinking about how to minimize measurement variation. That is, ideally, you should be able to get almost the same number of peanuts each time for each person. Describe what you did to minimize measurement variation.

2. Describe your procedure precisely, so that someone reading your description could replicate what you did.

3. Have each person gather as many peanuts she or he can hold in her or his hand.

4. Analyze and present your results. Include

   a. What you learned about minimizing measurement variation

   b. The analyses you did of your data: computations and graphs

   c. Your conclusions

NUMBER 4: How much did you improve while shooting hoops?

Materials

- Several identical waste baskets, balls (I recommend crumpling sheets of recycled notebook paper)

1. Take some time to standardize the procedure.
2. Each person takes 20 shots.
3. Analyze the data: What analyses help you to better understand the situation and better answer the questions?
   a. How did you do? How did the class do?
   b. Describe the variation among the class. How can you be more precise than saying "a little" or "a lot"?
4. Improvement
   a. Ask individuals with the best averages to describe what they did. Then take some practice time.
   b. Collect the new data.
   c. Analyze your new data, and answer these questions—how much did you improve, how much did the class improve?
5. Collect additional data and compare these results to previous results. Shoot from different distances, distract the shooter just like in pro games, place the waste basket in a corner, use a waste basket of different size.

NUMBER 5: Improvement in drawing angles

1. Make an angle that you think is 60 degrees using a straightedge or ruler.
2. With a protractor, record the measure of the angle.
3. On a separate sheet of paper (recycled if possible), make another angle that you think is 60 degrees.
4. Record the measure of the angle.
5. Your instructor will facilitate the collection of all the students' data, either on a piece of paper or on computer software.
6. Make a line plot of each set of data: your first and second attempts to draw a 60-degree angle.
   a. What can you conclude about the overall accuracy of students' first and second measurements?
   b. What can you conclude about how much the class as a whole improved?

7. Make a box plot of each set of data.

   a. What can you conclude about the overall accuracy of students' first and second measurements?

   b. What can you conclude about how much the class as a whole improved?

8. a. What could you see or conclude from the line plot that you didn't see or conclude from the box plot?

   b. What could you see or conclude from the box plot that your didn't see or conclude from the line plot?

9. a. What measures of the center were more useful in your analysis: mean, median, or mode (or modal class)? Explain.

   b. What tools for describing and understanding spread and dispersion—range, gaps, clusters, standard deviation—were useful? Explain.

10. *Extension:* Repeat this process with angles of different measures, for example, 45 degrees, 120 degrees, etc.

EXPLORATION 7.10   Collecting Data of Your Choice

1. Select and refine a question for which collection of data will provide some "answers." For example, "How much sleep do I get?" can be refined to "What time do I go to sleep and what time do I wake up?"

2. Decide how to collect and represent your data.

   a. What data will you collect?

   b. How will you collect and record the data?

   c. How precise will your "measurements" be? It will help at this time to look back and see whether the data you are collecting and how you are collecting the data will enable you to answer the question(s) you are asking.

   d. Think about how you will represent the data. For example, how will you graph them and how will you show measures of center and variation? (*Note:* You don't have to analyze all the data you collect. A student who chooses to collect data on her sleep might sum up Step 2 with "I will make a chart to record the day, my bedtime, and my waking time. I will round the times to the nearest five minutes."

3. Collect the data.

4. Analyze and represent the data. See the suggestions in Step 2(d).

5. Your instructor will specify the format for your report.

EXPLORATION 7.11 Designing and Conducting a Survey

In this exploration, you will design, conduct, and present results of a survey. My students report that this is both an interesting and a challenging project. In the first edition of the textbook, I did not have this exploration, because I knew that within the time constraints of this course, it is virtually impossible to determine a truly random sample. However, when I tried surveys with my class, I found that one of the major learnings from the exploration was an appreciation of the complexities of doing a survey and a more critical eye when reading and listening to the results of surveys. Because we are presented with survey results daily, in newspapers and on television, and because doing surveys is becoming increasingly common in elementary school textbooks, I decided to include a survey exploration.

1. Select a theme and at least four questions related to that theme. Carefully consider the questions that you will ask. You need to make sure that the respondents are answering the question you think you are asking! For example, when you ask whether they like the college, people define *like* in different ways. Similarly, you need to consider whether you want the question to be open-ended or forced-choice. For example, one or more groups in my class always collect data about alcohol use. If you ask how much alcohol someone has consumed in the past week, and some respondents write, "more than 10 drinks," this presents problems when you try to determine the average.

   You also need to determine whether you will ask respondents to fill out a questionnaire or to respond verbally. *Note:* When you ask questions that are sensitive, such as questions about alcohol consumption, you are more likely to get honest responses if the respondents feel some sense of anonymity. For example, you can have them fill out the survey, fold the paper, and place it into a box with other surveys.

   Essentially, you want to minimize the many problems that may occur in the data (such as missing data when someone completes only some questions and "bad" data when someone gives a response that can't be used or quantified).

2. Determine your target population. In most cases, your target population will be students at your college.

3. Devise, describe, and justify a strategy to get a representative sample. This discussion involves both logistics and justifications.

   a. Logistics: Where, when, and how will you collect the data?

   b. Justification: Why do you think the where, when, and how will give you a representative sample?

4. Determine how you and your group members will code the data. For example, if you want to compare males and females, then you have to code their responses separately.

5. Collect the data.

6. Analyze the data.

   a. First, analyze the data from your own sample. Determine the centers and variation for each question, and make sketches of appropriate graphs for each question.

   b. Compare results with other members of your group. Are your results (such as centers and variation) close or not?

7. Combine the data that you and your group members have collected. Determine what you have found and how best to present your findings to the class: centers and spreads, graphs, the text of your presentation, and the like.

8. Present your report to the class. Your instructor will specify the format for your report.

SECTION 7.3 Exploring Concepts Related to Chance

Chances are, you have some understanding of probability concepts. For instance, you know that when you roll two dice, the probability of getting a 2 is less than the probability of getting a 7. You know that the probability of snow is less in Atlanta than in Buffalo.

Like the concept of "average," "chance" also has a specific meaning that you will explore in this section. In the process, you will use some tools you have developed in earlier chapters and will develop some new ones.

Mathematics for Elementary School Teachers pp. 425, 432

EXPLORATION 7.12 Heads and Tails and Probability

The simple question of flipping a fair coin has captured the attention of people for many years. While we know that theoretically the probability of heads or tails is 50%, you have probably noticed that sometimes we see heads or tails appear several times in a row. This exploration will deepen your understanding of the concept of randomness. For many people randomness and haphazard are synonyms. However, as you will soon discover, while we cannot predict the outcome of a particular tossing of a coin, we can fairly accurately predict the outcome of a large number of cases. That is, there are patterns in randomness!

1. Flip a coin 10 times and record the number of heads.

2. Gather data from the entire class so that you have 50 samples of sample size 10.

3. Make a bar graph from the 50 samples.

   a. How often did you get exactly 5 heads and 5 tails?

   b. How would you describe the shape of the bar graph?

   c. What generalizations can you make from the bar graph?

4. Make at least two more bar graphs, each consisting of 50 samples. These graphs represent the same phenomenon, but the graphs are not the same.

   a. Why not?

   b. What generalizations can you make from observing these graphs?

5. Either by hand or with technology (using graphing calculators or a spreadsheet), construct an ongoing line graph of the cumulative probability of heads.

6. How would you describe what happens as you get more and more data?

7. Summarize what you learned from this exploration.

Strings

1. Let's say we tossed a coin 50 times. On average, what do you think will be the longest string of heads or tails when throwing a fair coin 50 times?

2. Gather sufficient samples to answer this question. If your instructor directs you, use a spreadsheet or other technology.

3. Make a bar graph from your data.

4. What is your response to this question? Support your conclusion.

Mathematics for Elementary School Teachers
pp. 424, 427

EXPLORATION 7.13   What Is the Probability of Having the Same Number of Boys and Girls?

In the children's story *Twenty-One Balloons,* by William Dubois, a man selects 25 families to join him on a remote, uninhabited island. However, there is a condition: He will only take families that have exactly 1 girl and 1 boy.

PART 1: Determining the probabilities experimentally

We will begin the exploration with a slightly more complex question.

1. What if a couple decided to have 4 children? What is the probability that they will have exactly 2 girls and 2 boys? Write your initial prediction and reasoning.

2. After the class discussion, revise your initial prediction and reasoning if you wish.

3. Simulate this problem. Collect samples of sample size 10. That is, how many times out of 10 do you get 2 boys and 2 girls? Do this a total of 50 times.

4. Make a line plot or bar graph of your data.

5. What do you think now?

6. How would you describe the distribution of the line plot or bar graph?

7. Using technology (graphing calculator or a spreadsheet), construct an ongoing line graph of the cumulative occurrence of 2 boys and 2 girls.

8. How would you describe what happens as you get more and more data?

9. What is the empirical probability of having 2 girls and 2 boys in a family of 4?

PART 2: Determining the probabilities theoretically

Now we will develop the theoretical probability and see some neat patterns!

1. Detemine the theoretical probabilities for the following:

    a. For a family of 2: 2 girls, 1 girls, 0 girls.

    b. For a family of 3: 3 girls, 2 girls, 1 girl, 0 girls.

    c. For a family of 4: 4 girls, 3 girls, 2 girls, 1 girl, 0 girls.

    d. For a family of 5: 5 girls, 4 girls, 3 girls, 2 girls, 1 girl, 0 girls.

2. Represent the result in a table like the one below. The second row has been filled out. That is, if you have two children, there is one way to have 0 girls, there are two ways to have 1 girl, and there is one way to have 2 girls, for a total of four possibilities. Describe pattern you see in this table.

		Number of girls						
		0	1	2	3	4	5	Total
Number of children	1							
	2	1	2	1				4
	3							
	4							
	5							

3. Use the table to determine the theoretical probability of having 3 girls and 3 boys in family of 6.

EXPLORATION 7.14   **What Is the Probability of Rolling Three Doubles in a Row?**

Almost everyone in the class has either played *Monopoly* or heard of it. In this game, if you roll doubles, you can roll again. However, if you roll three doubles in a row, you go to jail.

1.   What do you think is the probability of rolling three doubles in a row?

Let us work up to answering this question. We will assume that we have fair dice. Therefore, the probability of each number is equal.

2.   What if we roll two dice? What is the probability of rolling doubles? Write down your initial answer to this question and your reasoning.

3.   In many classes, we have two different answers. All students agree that there are six ways of rolling doubles. However, some students believe that the sample space consists of 21 outcomes and some believe the sample space consists of 36 outcomes. The former group considers (1, 2) and (2, 1) to be the same outcome and the latter group considers them to be two different outcomes. If this is an issue in your class, discuss this question until you come to resolution. There are two ways, both having merit, to resolving this question.

One involves a discussion of theoretical probability, that is, theoretically, what is the size of the sample space.

The other involves collecting data. If the first answer is correct, then, with more and more data, the probability will tend to converge to $\frac{6}{21}$ while if the second answer is correct, with more and more data, the probability will tend to converge to $\frac{1}{6}$.

Use either or both of these means to resolve the dilemma. If you originally had the wrong answer, take some time now to write what your thinking was that led you to the wrong answer and describe what helped you to understand the right answer.

4.   Now we move up one level in complexity. What is the probability of rolling two doubles in a row? Write your first thoughts and your justification of your thinking. How confident are you that your reasoning is valid? As before, discuss this question with the class. Again, you can gather empirical data.

5.   Now we move to the original question: What is the probability of rolling three doubles in a row? Write your answer and your justification.

6.   If you rolled three dice and added the numbers, what is the probability that you would get 13?

*Mathematics
for Elementary
School Teachers*
pp. 433, 438

EXPLORATION 7.15   What's in the Bag?

Your instructor will have placed a large number of objects in a bag (e.g., different color marbles). There are many of these in the bag, and there are two colors. By sampling, you are to determine the proportion (percentage) of each.

1. Take out 10 objects and record the number of the designated color.

2. Once you get 10 samples of sample size 10, make a line plot.
    Looking only at the distribution, what do you see?

3. Collect more data until you have 50 samples of sample size 10.

    Make a line plot for these data.
    Looking only at the distribution, what do you see?

4. Repeat step 3 to generate data for four line plots, each containing 50 samples of sample size 10.
    Describe similarities and differences between the various distributions.

5. Summarize what you have learned from this exploration about randomness, about patterns in randomness, and about what is the same and what is different when you look at different sets of data from the same bag.

Assessment:

6. Students drew samples from a bag that had 1000 marbles. The bag had blue and red marbles. Each of four groups collected 50 samples of sample size 10. On the next page are the line plots from each group showing the number of blue marbles in each sample of 10. Just by looking at the four line plots, state what you think is the proportion of each color and justify your results.

```
                            X
                            X
                            X
                            X
                            X
                        X   X
                        X   X
                        X   X
                        X   X
                        X   X
                    X   X   X   X
                    X   X   X   X
                    X   X   X   X
            X       X   X   X   X   X
            X       X   X   X   X   X
            X   X   X   X   X   X   X
 0   1   2   3   4   5   6   7   8   9   10
                  Graph A
```

```
                            X
                            X
                            X
                            X   X
                            X   X
                            X   X
                            X   X
                    X       X   X
                    X       X   X
                    X   X   X   X   X
                    X   X   X   X   X
                    X   X   X   X   X
                X   X   X   X   X   X
     X          X   X   X   X   X   X
 0   1   2   3   4   5   6   7   8   9   10
                  Graph B
```

```
                            X
                            X
                        X       X   X
                        X   X   X   X
                        X   X   X   X
                        X   X   X   X
                        X   X   X   X
                    X   X   X   X
                    X   X   X   X   X
                    X   X   X   X   X
                    X   X   X   X   X
                    X   X   X   X   X
 0   1   2   3   4   5   6   7   8   9   10
                  Graph C
```

```
                            X
                        X   X
                        X   X
                    X   X   X
                    X   X   X   X
                    X   X   X   X
                    X   X   X   X
                    X   X   X   X
                    X   X   X   X   X
            X       X   X   X   X   X
            X   X   X   X   X   X   X
 0   1   2   3   4   5   6   7   8   9   10
                  Graph D
```

EXPLORATION 7.16   **How Many Boxes Will You Probably Have to Buy?**

Cereal companies often place prizes inside boxes of cereal to attract customers. One of these promotions lends itself nicely to exploration. Many elementary school teachers have done this exploration, in a simpler form, with their students. It certainly captures their attention!

Inside specially marked packages of Sugar Sugar cereal, you can get a photograph of one of six sports figures. Many kids will want to have one of each.

1. If we want to get all six prizes, how many boxes do you predict we will have to buy, on average?

    a. Write your prediction and justify your reasoning.

    b. (Optional) Gather everyone's prediction and summarize what the class prediction is—average and variation, distribution.

2. Each group simulates the situation 10 times.

3. **a.** On the basis of 10 simulations, what do you predict now? Justify your reasoning.

    b. What have you learned or noticed from these 10 turns?

4. Collect the data from the whole class.

    a. Analyze those data. What do you predict now?

    b. Indicate which operations on the data were useful—computations of mean, median, mode, range, standard deviation, etc. Which graphs were useful?

*Mathematics
for Elementary
School Teachers*
p. 427

EXPLORATION 7.17   More Simulations

The following situations present real-life questions for which determining the theoretical probability is either tedious or impossible. Thus they lend themselves to simulations in which we can determine the experimental probability.

For each of the following questions, develop and execute a simulation plan to help you answer the question. Your instructor will specify the format for your reports.

1. *The basketball game is on the line*   Horace has just been fouled, and there is no time on the clock. His team is down by one point, and he gets two shots. Horace is an 80% free-throw shooter. What percent chance of winning in regulation time does his team have?

2. *Genetics*   Maggie and Tony have just discovered that they are carriers of a genetically transmitted disease and that they have a 25% chance of passing on this disease to their children. They had planned to have four children. If they do have four children, what is the probability that at least one of the children will get the disease?

3. *Overbooking*   Airlines commonly overbook; that is, they sell more tickets for a flight than there are seats.

    a. Why do you think airlines overbook?

    b. An airline has a plane with 40 seats. What information will help it to decide how much to overbook?

    c. Using the information from your instructor, develop a simulation plan to help the airline decide how many tickets to sell. Meet with another group to share ideas. Note any changes in your plan.

    d. Do the simulation. What is your conclusion? Support your conclusion.

4. *Having children*   In an attempt to deal with overpopulation, the Chinese government has a policy that couples may have only one child. One of the biggest problems with this policy is that in Chinese culture, having a boy is preferable. As a result of the policy and the preference for boys, infanticide is not uncommon in China (and in other countries too—for example, in parts of Nepal, where I served in the Peace Corps). That is, it is not uncommon for parents to kill their newborn baby if it is a girl. Obviously, there are many ways to deal with this problem.

    Two questions that arise from this policy are (1) how would it affect family size, and (2) would it affect the ratio of boys to girls? Assume that this policy was implemented. Develop and run a simulation to answer these questions.

EXPLORATION 7.18   **Using Sampling to Estimate a Whole Population**

This exploration is a simulation of a method that has been used by naturalists to estimate the number of fish in a lake.

Your group will obtain a supply of "fish" from your instructor, and you will estimate the total number of fish without counting every single one.

1. Obtain your "fish" from your instructor and place them in a container. The container represents the lake in which these fish are distributed. Imagine that there were concerns that the lake's fish population was declining and that this would hurt tourism, which in turn would hurt the local economy. You have been asked to determine the number of fish in the lake. This number will be used as a baseline so that, when the procedure is repeated in succeeding years, analysts will be able to see whether the population is actually declining.

    a. Brainstorm possible ways to estimate the number of fish in the lake by a means other than counting all of them.

    b. Select one method and try it out. Write up this method, your justification of the method, and an estimate of the number of fish you got when you applied this method.

2. Now we will explore a method that scientists have used. First, they catch a "reasonable" number of fish and then tag and release them. They wait some time. Then they catch another batch of fish; some of these fish will have tags and some will not. We now have enough data to estimate the number of fish in the lake. Discuss this procedure in your group until you understand both how and why it works. You may want to do a trial run to help you better understand the procedure.

3. Now use the procedure with the fish that your instructor gave you.

    a. Decide how many fish to tag and briefly justify your choice. Tag that number of fish.

    b. Determine the size of your sample and briefly justify your choice. Catch the fish and record the number of caught fish that are tagged and the number that are untagged.

    c. Use this information to estimate the total number of fish in the lake.

    d. Repeat parts (b) and (c) until you feel that your estimate is "close." Briefly explain why you think your estimate is close. You may want to use a table like the following:

Sample	Number caught	Number caught that were tagged	Estimate of actual population
1			
2			
3			

4. Have a class discussion in which each group presents its estimate, justifies its choice of number of fish to tag and number of fish to catch again, and explains its degree of confidence that its estimate is close to the actual population. What did you learn from the class discussion?

5. Imagine doing this in a real lake with real fish. What factors might cause the tagged fish not to be randomly distributed in the lake? How might the biologists deal with these issues?

Mathematics for Elementary School Teachers p. 440

EXPLORATION 7.19 Fair Games

One exercise commonly used in elementary schools to help children learn important probability ideas is determining whether a game is fair or not. For example, if you flip a coin, specifying that player A gets a point if the coin lands on heads and player B gets a point if the coin lands on tails, this is a fair game because both outcomes are equally likely. We will begin with some simple games and then move to more complex games.

Here are the directions for each game:

1. Predict whether you think this is a fair game or not. Explain your response.

2. Play the game as many times as specified by the instructor.

3. Now do you think the game is fair or not? Justify your response. If you think the game is not fair, explain how to change the game to make it fair, and justify your change.

Coin Games

1. The game: Flip two coins.

   Player 1 gets a point if the coins match.

   Player 2 gets a point if the coins don't match.

2. The game: Flip three coins.

   Player 1 gets a point if the coins are all the same or all different.

   Player 2 gets a point otherwise.

Dice Games

1. The game: Roll two dice and add the numbers.

   Player 1 wins if the sum is even.

   Player 2 wins if the sum is odd.

2. The game: Roll two dice and add the numbers.

   Player 1 gets 3 points if the sum for the dice is seven.

   Player 2 gets 1 point otherwise.

3. The game: Roll two dice and multiply the numbers.

   Player 1 wins if the sum is even.

   Player 2 wins if the sum is odd.

4. The game: Roll two tetrahedral dice and add the numbers.

   Player 1 wins if the sum is even.

   Player 2 wins if the sum is odd.

5. The game: Roll three dice and add the numbers.

   Player 1 wins if the sum is even.

   Player 2 wins if the sum is odd.

Spinner Games

1. The game: Two spinners, each is half one color and half another color. Spin both spinners.

   Player 1 wins if the colors are the same on both spinners.

   Player 2 wins if the colors are different.

2. The game: Two spinners, each divided into thirds, one color on each third. Spin both spinners.

Player 1 wins if the colors are the same on both spinners.

Player 2 wins if the colors are different.

Scissors, Rock, Paper

Three people play. At the count of three, each person displays scissors, rock, or paper.

Player 1 gets a point if all three are the same.

Player 2 gets a point if all three are different.

Player 3 gets a point otherwise.

Marshmallow

The game: Throw a marshmallow.

Player 1 wins if it lands on an end.

Player 2 wins if it lands on its side.

SECTION 7.4   Exploring Counting and Chance

Many real-life questions and problems fall into the field of probability called *combinatorics*. That is, there are patterns that enable us to determine the number of elements in the sample space and the number of elements in the subset in which we are interested.

Mathematics for Elementary School Teachers p. 447

EXPLORATION 7.20   License Plates

When cars first started being sold, there were no license plates and no driver's licenses. If you could afford a car, you could drive it! However, as cars became more popular, the need to require people to register their cars arose.

Many elementary and middle school teachers have found the origin of license plates an extremely interesting topic to explore with their students. It turns out that different states made different decisions about what to do when they ran out of combinations.

1. **M 728**

    a. An early license plate configuration consisted of a letter of the alphabet followed by 3 digits. With this system, how many possible license plates could a state make?

    b. A next step would be simply to add a digit—that is, to use a letter followed by 4 digits. How many different license plates could be made this way?

    c. Let's say that your state uses the system described in part (b) and is about to run out of license plates. What would you recommend next? Why?

2. **123 456**

    a. At some point, most states have used the following system: License plates consisting of a 6-digit number. How many possible license plates are there in this system?

    b. Let's say your state has been using the system described in part (a) and is about to run out of license plates. What would you recommend next? Why?

3. **3 G 2346**

    a. Some states went for the pattern shown here: a digit, then a letter, followed by 4 more digits. How many possible license plates are there in this system?

    b. Many states discovered that 3 letters and 3 digits make a sufficient number of combinations. How many possible license plates are there in this system?

4. Let's say a state outgrew the license plate consisting of 3 letters and 3 digits and was considering these two options: 1 letter, 2 digits, and 3 letters, or 6 letters. Which would you recommend? Why?

5. Let's say a state decided to make a license plate that consisted of 6 letters. Theoretically, this would make $26 \times 26 \times 26 \times 26 \times 26 \times 26$ possible license plates. However, Mason has a problem. His calculator doesn't have enough spaces to show the exact answer; it displays only 8 digits. It shows 3.0891578 08. Explain how you could determine the actual answer with a *minimum* of pencil and paper computation.

6. Let's say a state decided to have a license plate with the following format: 2 letters followed by 3 digits followed by 1 letter—for example, AB123C.

    a. How many different license plates could be made with this system?

    b. Maegen says that the answer here is the same as that for 3 letters followed by 3 digits. What do you think? Justify your answer.

Mathematics for Elementary School Teachers
p. 447

EXPLORATION 7.21 Native American Games

Human beings have played games of chance for thousands (perhaps tens of thousands) of years. Stewart Culin wrote a book called *Games of the North American Indians.*[1] This exploration uses some of the games in Culin's book and will help you to grapple with basic probability concepts, have some fun, and gain a sense of history.

Directions (for each part)

1. *Predictions*   After playing for only a few throws, predict the answers to the following questions.
   a. What is the probability that you will score at least 1 point on your turn?
   b. What is the probability that you will score no points on your turn?
   c. How many different outcomes are there?

2. *Experimental probability*   Play the game for some time and determine answers to the first two questions in Step 1. That is, determine the experimental probabilities.

3. *Theoretical probability*   Determine the theoretical probabilities for all three questions in Step 1.

4. *Presentations*   In your presentation, you must
   a. Demonstrate your game.
   b. Give your answers to the three questions in Step 1, making sure that the listener can understand how you got your answers. You are free to make tables, charts, diagrams, and whatever other visual aids you consider helpful.
   c. Share any other insights or questions about the game.

   In each case, you may make your own sticks or select other materials that will create an equivalent game.

PART 1: A game from the Klamath tribe in California[2]

Four sticks are marked as shown at the right. The two top sticks are called *shnawedsh* (women), and the two bottom sticks are called *xoxsha* or *hishuaksk* (men). The lines were made by pressing a hot, sharp-pointed tool against the wood. The other side of each stick is unmarked (plain).

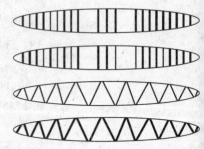

   The game consists of throwing the four sticks. The scoring is as follows:

- If all four marked sides face up, 2 points.
- If all four plain sides face up, 2 points.
- If both male sticks are up and the female sticks down, 1 point.
- If both female sticks are up and the male sticks down, 1 point.

PART 2: A game from the Nishinam tribe in California[3]

Two acorns have been split lengthwise in halves, and the outsides have been painted red or black. The game consists of throwing the four acorn halves. The scoring is as follows:

- If all four painted sides face up, 4 points.
- If all four painted sides face down, 4 points.
- If three painted sides face up, no points.
- If two painted sides face up, 1 point.
- If one painted side faces up, no points.

In this game, a player keeps throwing until a throw results in 0 points, and then it is the next player's turn.

PART 3: A game from the Songish tribe from British Columbia[4]

Four beaver teeth have been marked in the following manner on one side:

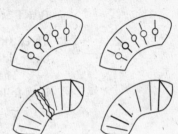

- Two teeth have been marked with a row of circles; they are called *women*.
- The other two teeth have been marked with cross lines; they are called *men*.
- One of the male teeth is tied in the middle with a small string and is called *ihkakesen*.

The game consists of throwing the four teeth. The scoring is as follows:

- If all four marked sides face up, 2 points.
- If all four marked sides face down, 2 points.
- If both male teeth are up and the female teeth are down, 1 point.
- If both female teeth are up and the male teeth are down, 1 point.
- If the ihkakesen is face up and the other teeth face down, 4 points.
- If the ihkakesen is face down and the other teeth face up, 4 points.

In this game, a player keeps throwing until a throw results in 0 points, and then it is the next player's turn.

PART 4: A game from the Zuni tribe in New Mexico[5]

Four sticks have been painted red on one side; the other side is plain.

The game consists of throwing the four sticks. The scoring is as follows:

- If all four painted sides face up, 4 points.
- If three painted sides face up, 3 points.
- If two painted sides face up, 2 points.
- If one painted side faces up, 1 point.

In this game, a player keeps throwing until a throw results in 0 points, and then it is the next player's turn.

PART 5: Another game from the Zuni tribe in New Mexico[6]

Three sticks have been painted red on one side and black on the other side.

The game consists of throwing the three sticks. The scoring is as follows:

- If the three red sides face up, 10 points.
- If the three black sides face up, 5 points.
- If two red sides and one black side face up, 3 points.
- If one red side and two black sides face up, 2 points.

In this game, if three red sides face up, the thrower gets another turn.

Geometry as Shape

E xplorations 8.1, 8.2, and 8.3 make use of three types of manipulatives—Geoboards, tangrams, and polyominoes—that are helpful in developing a strong understanding of many important mathematical ideas. These explorations have multiple parts and address topics from throughout Chapter 8. Your instructor may choose to use parts of these explorations in combination with other explorations from the individual sections, which follow these opening explorations.

Mathematics for Elementary School Teachers
p. 465

EXPLORATION 8.1   Geoboard Explorations

Geoboards are a popular manipulative in elementary schools and are very versatile. Smaller Geoboards generally contain 25 pegs (5 rows of 5 in a row), and larger Geoboards generally contain 100 pegs. There are also circular Geoboards, generally with 24 pegs arranged in a circle.

PART 1: Communication

This first part of the exploration both serves as an introduction to Geoboards and reinforces the need for clear communication when talking about shapes.

Instructions: In this exploration, you will work in groups of 3.

Persons A and B each have a Geoboard (or Geoboard Dot Paper) and sit back to back. Person C is the observer.

Person A makes a figure on the Geoboard. Next, using only words, person A gives directions so that person B can construct the same figure on his or her Geoboard. The observer watches without comment and will give feedback at the end. Person B's responsibility is to ask for clarification whenever person A's directions are not clear.

1. After you are finished, compare Geoboards. The figures may or may not be identical. In either case, listen to person C's feedback and discuss how to make communication easier and pinpoint places where communication broke down.

2. Rotate roles: Person C now makes the figure and gives directions, person A makes the copy, and person B is the observer.

3. Repeat this process once more so that each person has a turn in each role.

4. Afterwards, describe what ideas or terms you learned to make communication easier.

PART 2: Challenges

You may use your Geoboards to solve the problems below. Afterwards, transfer your answer to Geoboard Dot Paper (at the back of this book).

Children often say, "I wonder if . . ."—for example, "I wonder if you can have a figure with exactly 3 right angles." "I wonder how many different figures you can have that have only one peg inside." Pursuing these questions can help you develop problem-solving and reasoning skills and also help you to understand some of the properties of geometric figures.

1. In this problem, construct the following figures on your Geoboard (or Geoboard Dot Paper) or explain why you think it is impossible to do so.

    a. A figure with just 1 right angle

    b. A figure with 2 right angles

    c. A figure with at least 1 right angle but no sides parallel to the edges of the Geoboard

    d. A figure with 6 right angles

    e. A figure with exactly 2 congruent, adjacent sides

2. Find all possible squares that can be made on a 25-peg Geoboard.

3. **a.** Make a quadrilateral with no parallel sides.

    b. Make a parallelogram with no sides parallel to the edge of the Geoboard.

    c. Make two shapes that have the same shape but are different sizes.

    d. Make two polygons that have different shapes but the same area.

4. **a.** Make a pentagon. Then list all the attributes of this pentagon that you can.

    b. Make another pentagon and list all the attributes of this pentagon that you can.

    c. Now list the characteristics they have in common.

PART 3: Making triangles and quadrilaterals

Problem 1 appears in the "Navigating Through Geometry in Grades 3–5" book. You can examine the children's work after the exploration.

1. How many different right triangles can you make on a 5 × 5 Geoboard? There is a lot to think about here:

    - Can you think systematically, as opposed to just randomly?
    - How will you represent your triangles on the paper so you can keep track of them?
    - How do you know an angle is a right angle? There are right triangles whose bases are not parallel to the edges of the paper!

    On the paper you turn in, include

    a. your triangles and a brief description of why you placed the triangles in the order you did

    b. your method(s) for systematically determining all the triangles

    c. how you identified right angles whose sides were not parallel to the edges of the paper. Here we are looking for more than "it looked like a right angle" or "I used a protractor."

Problem 2 was posed in the October 2000 issue of *Teaching Children Mathematics*. The solution, containing results submitted by elementary school teachers at various grade levels, appears in the October 2001 issue.

2.   How many different triangles can you make on a 5 × 5 Geoboard that have no pegs in the interior of the triangle? Refer to Problem 1 for further discussion and hints. On the paper you turn in, include

   a. your triangles and a brief description of why you placed the triangles in the order you did

   b. your method(s) for systematically determining the triangles

3.   How many different triangles can you make on a 5 × 5 Geoboard? Refer to Problem 1 for further discussion and hints. On the paper you turn in, include

   a. your triangles and a brief description of why you placed the triangles in the order you did

   b. your method(s) for systematically determining the triangles

Your instructor might choose to make the problem slightly smaller by specifying 4 × 4 instead of 5 × 5 Geoboards. This problem is substantially more complex than the previous two. You really need to think about being systematic, and you also need to think about how you will represent your answer. That is, it makes it easier to compare answers if there is some kind of organization of your triangles. Some students have found it useful to invent their own notation for triangles, such as 1 × 2, 1 × 3, etc.

PART 4: Recognizing and classifying figures with Geoboards

Geoboards can also be used to help students appreciate the need for names for various geometric figures and geometric ideas. In elementary, middle, and high school, you learned the names and properties of a number of shapes, such as isosceles triangle, parallelogram, and regular polygon. For example, there are (infinitely) many isosceles triangles, but regardless of their size and angles, all isosceles triangles have certain attributes in common. In this exploration, you explore different kinds of subsets from a small universe of geometric figures.

1.   **a.** Cut out the set of quadrilaterals on page 189.

   b. Separate them into two or more groups so that the members of each group are alike in some way and so that each quadrilateral belongs to exactly one group.

   c. Describe each group so that someone who couldn't see your figures could get a good idea of what each group looked like.

   d. Name each group. If you know of a mathematical name for one or more groups, use it. If you don't know of a mathematical name, make up a descriptive name that fits the group.

2.   Repeat this process as many times as you can in the time given. That is, in what other ways can we separate these figures into two or more groups so that every group has one or more common characteristics?

3.   After the class discussion, answer the following questions:

   a. What did you learn from this exploration?

   b. How do definitions help us to communicate when discussing geometry?

Figures for EXPLORATION 8.1: PART 4, Recognizing and classifying figures with Geoboards

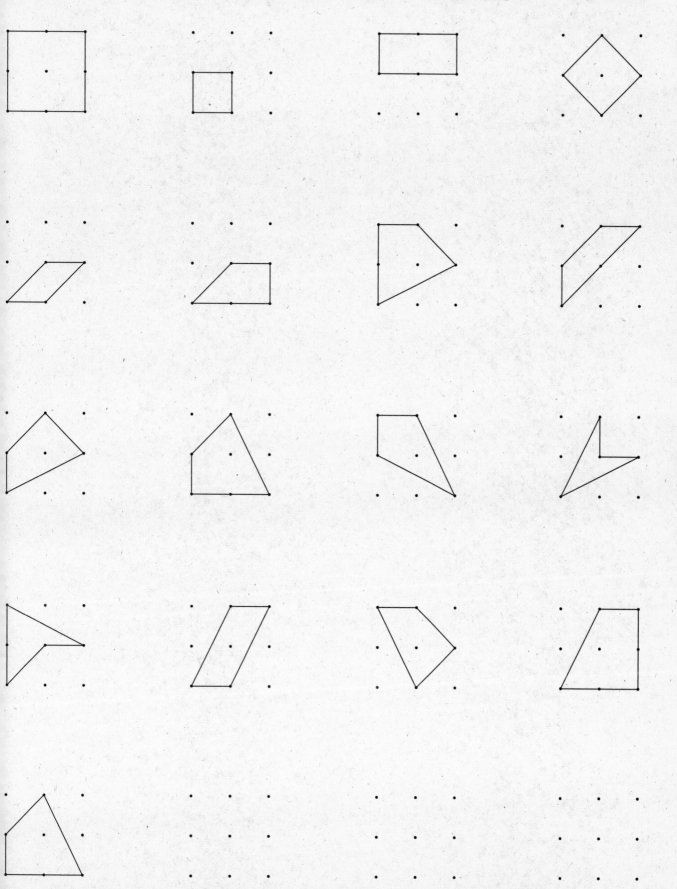

Mathematics
for Elementary
School Teachers
p. 465

EXPLORATION 8.2   Tangram Explorations

Tangrams are another very versatile manipulative that we will use on several occasions in this and the next two chapters. Tangrams were invented in China at least two hundred years ago, but we are not sure by whom or what for. They quickly became a popular puzzle, because there are so many different things you can do with them! The English word *tangram* probably came from American sailors who referred to all things Chinese as Tang, from the Cantonese word for China.

You will use tangrams provided by your instructor or make your own set from the template at the back of the book.

PART 1: Observations, discoveries, and questions

The directions here are very simple: Play with the tangrams for a while. As you explore the shapes, you will notice things: patterns and relationships among the various pieces. You might make interesting shapes. You may ask yourself various "what if" questions.

Record your observations: patterns, discoveries, conjectures, and questions.

PART 2: Puzzles

As you may have found, you can make a number of interesting shapes with tangrams. Below are some famous puzzles that call for careful thinking.

1.   **a.** Use all seven tangram pieces to make each of the five figures shown on page 193. Sketch your solution.

   b. Describe any thinking tools that you became aware of while figuring out how to make the figures.

   c. Describe any new observations: patterns, discoveries, conjectures, and questions.

2.   Make your own puzzle. Name it. Give it to someone else to solve.

3.   Make a figure with the tangram pieces. Write directions for making that figure, as though you were talking on the phone to a friend. Exchange directions with a partner. Try to make the figure from your partner's description.

   a. Discuss any problems that either of you encountered in understanding what the other had written.

   b. Now make a new figure and write directions for making that figure, as though you were talking on the phone to a friend.

Figures for EXPLORATION 8.2, PART 2: Puzzles

1.

PART 3: Combining and subdividing shapes

The following exploration is a variation of one I have seen in many places. It is commonly used in elementary school.

Fill in the table below. For example, the first cell is solved by putting the two small triangles together. There is another way to make a square with two pieces. Can you find it?

If there is more than one answer for a given cell, give it. If a given cell is impossible, explain why you think so.

Number of pieces	Square	Rectangle	Triangle	Right trapezoid	Isosceles trapezoid	Parallelogram
2						
3						
4						
5						
6						
7						
8						

PART 4: Classifying and naming geometric figures with tangrams

This exploration uses tangrams to explore names and properties of different shapes.

1.   *The two small triangles and the medium triangle*   Using only the two small triangles and the medium triangle, make as many shapes as you can, with the following restriction: When two pieces meet, the whole side of one piece must fit against the whole side of another piece. Record your findings on a separate piece of paper by tracing the shapes. Try to be systematic to ensure that you find all possible shapes.

a. Describe your strategies (other than random trial and error) for finding all possible shapes. Try to explain your strategies so that someone reading this description via e-mail could understand it.

b. Name each shape. If you know of a mathematical name, use it. If not, make up a name that is related in some way to the shape.

c. Describe your other observations from this exploration: patterns, discoveries, conjectures, and questions.

2. *The two small triangles and the square*   With the same restrictions, make as many shapes as you can using only the two small triangles and the square. Record your findings on a separate piece of paper by tracing.

a. Describe your strategies (other than random trial and error) for finding all possible shapes. Try to explain your strategies so that someone reading this description via e-mail could understand it.

b. Name each shape. If you know of a mathematical name, use it. If not, make up a name that is related in some way to the shape.

c. Describe your other observations from this exploration: patterns, discoveries, conjectures, and questions.

3. *The two small triangles and the parallelogram*   With the same restrictions, make as many shapes as you can using only the two small triangles and the parallelogram. Record your findings on a separate piece of paper by tracing.

a. Describe your strategies (other than random trial and error) for finding all possible shapes. Try to explain your strategies so that someone reading this description via e-mail could understand it.

b. Name each shape. If you know of a mathematical name, use it. If not, make up a name that is related in some way to the shape.

c. Describe your other observations from this exploration: patterns, discoveries, conjectures, and questions.

4. *Making connections*

a. Your instructor will have facilitated the making of a table that shows the shapes made in Steps 1, 2, and 3. In some cases all three columns are filled, indicating that this shape (for example, the "same" rectangle) could be made in all three cases. In some cases there will be blanks, indicating that the group was not able to make a particular shape with the three tangram pieces they had. Your task here is to examine the blanks that you or your instructor select and either make the shape or prove that it cannot be made.

b. The number of shapes that were made from the two small triangles and the parallelogram was greater than the number of shapes that could be made from the two small triangles and the medium triangle or from the two small triangles and the square. Explain why we get more shapes with the two small triangles and the parallelogram.

Mathematics for Elementary School Teachers
p. 465

EXPLORATION 8.3   Polyomino Explorations

Polyominoes are geometric figures that are composed entirely of squares. You are probably familiar with two subsets of polyominoes: *dominoes* (made from two squares) and *tetrominoes* (made with four squares and made most popular by the Game Boy game Tetris®). The word *polyomino* was coined by Solomon Golumb, an American mathematician, in 1953, though puzzles with different kinds of polyominoes seem to have been around for centuries. You will find explorations with polyominoes in many elementary school textbooks, because they are such a versatile tool for exploring different mathematical ideas and developing spatial sense and problem-solving skills.

PART 1: Defining pentominoes

We will begin our explorations with *pentominoes* and communication. Below is a picture of two figures that are pentominoes and two figures that are not pentominoes.

Pentominoes

Not pentominoes

1. Write down your definition of *pentomino*.

2. Exchange your definition with another student or read your definition to your group. Considering each definition as a first draft, identify any confusing or ambiguous words or phrases.

3. Write your second draft of your definition of *pentomino*.

PART 2: How many polyominoes?

In this exploration, you will create two sets of polyominoes—tetrominoes and pentominoes—using two or three pages of the Polyomino Grid Paper provided at the end of the book. Save your tetromino and pentomino sets for use in other explorations in Chapters 8 and 9.

All dominoes have the same shape. If you look at the two dominoes below at the left, you can see that the second domino is the same shape as the first; it has simply been rotated 90 degrees. When we get to *trominoes,* which are made with three squares, the matter is still rather simple. There are two different trominoes, as shown in the figure below at the right.

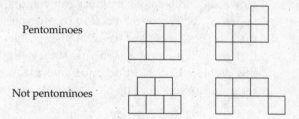

Dominoes   Trominoes

1. When we get to tetrominoes, made with four squares and popularized by the game Tetris®, there are a few more. Determine how many different tetrominoes there are. Show your solutions by cutting out your different tetrominoes from the Polyomino Grid Paper at the end of the book.

2. When we search for all tetrominoes, a more or less random trial-and-error strategy is often sufficient. However, when we move to pentominoes, the number of possibilities is great enough so that it pays to be more systematic and less random.

    a. How many "different" pentominoes can be made? Show your solutions by cutting out your different pentominoes from the Polyomino Grid Paper at the end of the book. Describe any tools other than random trial and error that you used in your search.

    b. Once you think you have found all of the pentominoes, explain why you think you have them all.

    c. As a class, come up with a name for each different pentomino. This is necessary to make communication easier.

3. How many different hexominoes can you find? (You can use a copy of "Other Base Graph Paper" at the end of the book to record your examples of hexominoes.) Because this is a mathematics course and we want your problem-solving tools to develop, the expectation is that you will use problem-solving strategies other than random trial and error. Think about this problem first. Next describe your plan of attack. Then do the problem, thinking and reflecting as you work on it. At the end, summarize your strategies.

PART 3: Classifying pentominoes

1. Take your set of pentominoes and divide them into two or more subsets so that the members of each subset are alike in some way and each pentomino belongs to exactly one group.

2. List the members of each subset.

3. Define the mathematical attributes of all members of each subset.

4. Repeat this as many times as your instructor asks.

PART 4: A pentomino game

Materials

- Your set of pentominoes (made in Part 2)
- Use a copy of the Polyomino Grid Paper at the end of the book to make a chessboard (8 × 8 square). Each square of your pentominoes is the same size as each of the 64 squares on the board.

Rules for competitive version

- Players take turns placing one pentomino on the board.
- The winner is the last person who can place a pentomino on the chessboard so that it does not extend beyond the board and does not lie on top of a previously played pentomino.

1. Play the game several times. Describe any strategies you discovered.

2. Play a cooperative version: Find the set of moves that results in the smallest number of pentominoes being placed on the board.

    a. Show your solution.

    b. Describe any strategies you discovered.

PART 5: Challenges

Use the tetromino and pentomino sets that you made in Part 2.
Note: In each of the challenges, describe strategies you used beyond guess–check–revise.

1. Using all five tetrominoes, completely fill grids (a) and (b) on page 201.

2. Using four different pentominoes, fill grids a and b on page 202.

3. Selecting among your set of pentominoes, solve the following puzzles.

   a. Use 4 different pentominoes to make a 4 × 5 rectangle.

   b. Use 6 different pentominoes to make a 3 × 10 rectangle.

   c. Use 5 different pentominoes to make a 5 × 5 square.

4. Using your whole set of 12 pentominoes, arrange the pentominoes so that the space enclosed by the pentominoes is as large as possible.

5. This challenge comes from a story, probably fiction, that goes like this: The son of William the Conqueror and the dauphin of France were playing a game of chess. At one point, the dauphin became angry and threw the chessboard at William's son. The board broke into thirteen pieces, 12 different pentominoes and 1 tetromino. Make a chessboard with the 12 different pentominoes and 1 tetromino.

6. Make up your own challenge. Describe your thinking process and a solution.

PART 6: Questions about categories

1. Juan sorted the pentominoes into these four groups. Write the directions for sorting so that the reader of this description will know how to put the pentominoes into the correct groups.

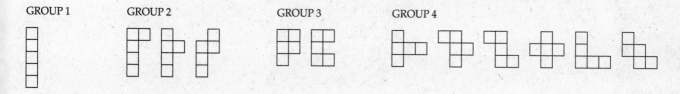

2. Lakela sorted the pentominoes into these three groups. Write the directions for sorting so that the reader of this description will know how to put the pentominoes into the correct groups.

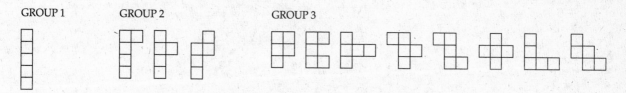

3. Ari sorted the pentominoes into these four groups. Write the directions for sorting so that the reader of this description will know how to put the pentominoes into the correct groups.

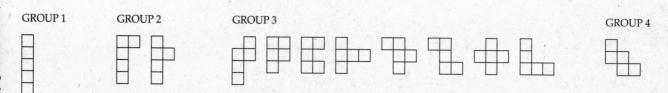

Grids for EXPLORATION 8.3, PART 5: Challenges

1.   a.

   b.

Grids for EXPLORATION 8.3, PART 5: Challenges

2.   a.

b.

Grids for EXPLORATION 8.3, PART 5: Challenges

SECTION 8.1   Exploring Basic Concepts of Geometry

In order to work effectively with geometric ideas, we need to have a foundation of postulates, definitions, and theorems. As you may recall from high school geometry, the Greek contribution to geometry and mathematics is enormous. Although this textbook does not attempt to replicate the kind of formal geometric work you did in high school, I do believe it is helpful for elementary teachers to explore concepts and ideas that students will further examine in high school in order to have a better sense of the connection between the explorations students do in elementary school and the geometric knowledge they need to bring to middle school and high school.

Mathematics for Elementary School Teachers pp. 463, 466

EXPLORATION 8.4   Manhole Covers

Geometry serves many purposes in our world. In this exploration, we will look at the geometric reason for the shape of one object.

PART 1: Manhole covers

1. Why do you think manhole covers are round? Write down what you think.

2. In order to understand why manhole covers are round, we will cut out a variety of shapes.

    a. Trace the circle and the square on pages 205 and 206 onto a blank sheet of paper, some space apart. Cut out the circle and the square in such a way that you only cut on the lines of the square and the circle. That is, fold the square in half and cut around the edges; do the same with the circle.

    b. You now have models of a circular and a square manhole cover. What do you notice?

    c. If you hold the square on edge and turn it, the length of the diagonal of the square is significantly longer than the length of one side, so the square can easily fall into the hole. The longest line segment you can draw across the hole made by the circle is equal to the diameter of the circle. How do you think the designers of manhole covers addressed this so that the circular manhole cover would not fall into the hole? Write your thoughts.

    d. How big a lip would you have to make for a square manhole cover so that the cover could not accidentally fall through?

PART 2: Cutting out shapes with the least number of cuts.

1. Trace the square on page 205 to another sheet of paper. Then fold the paper so that the fold cuts the square in half. Now you can cut the paper beginning at A and making your way around the square. To cut the square out while keeping the sheet of paper intact, we made use of some properties of a square: We know that a square has symmetry, so if we fold it in half, the top half will fit exactly onto the bottom half. Thus we need only cut half of the square to cut the whole square out of the paper. Looking a little closer, we see that the properties of a square that make it easier to cut by folding are that the opposite sides of the square are equal (congruent) and that all four angles are also equal (congruent).

What we just did counts as three cuts—from A to B, from B to C, and from C to D. There are two ways to fold the paper so that you can cut out the square with one cut. Try to figure how to cut the square with only one cut. Write down your solution. If you can only get it down to 2 cuts, write down your solution.

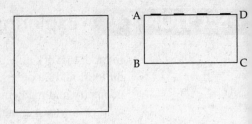

2. Trace each of the polygons on the page to another sheet of paper. Figure out how to cut the shape out with the least number of cuts. Before you do so, think of your strategy and the properties of the shape that enable you to make fewer cuts than the number of sides. Once you have your solution:

    a. Describe your strategy so someone could read your description and be able to do it.

    b. Describe the properties of the figure that enabled you to solve the problem.

Figures for EXPLORATION 8.4, PART 1: Manhole Covers

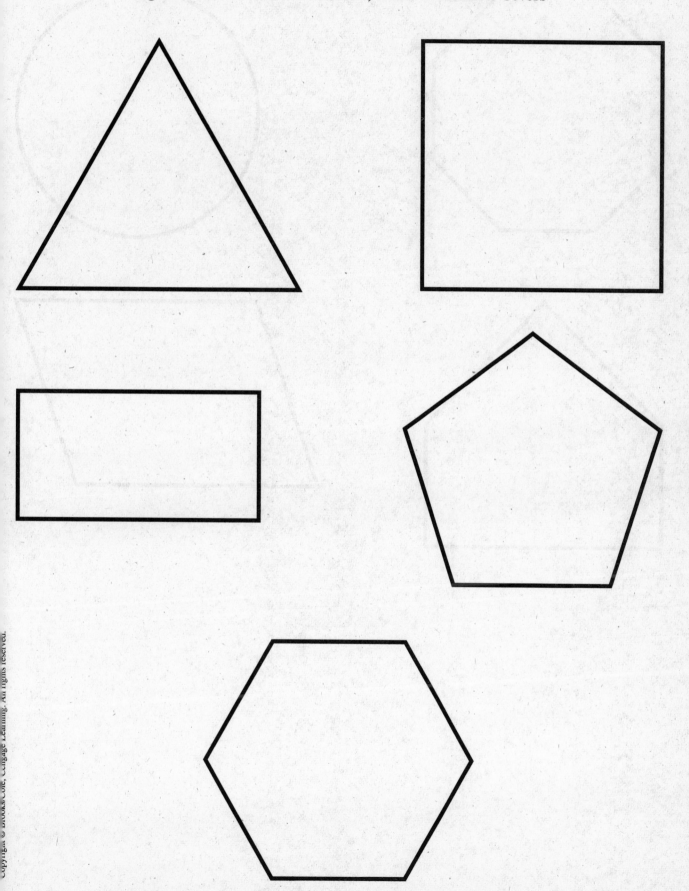

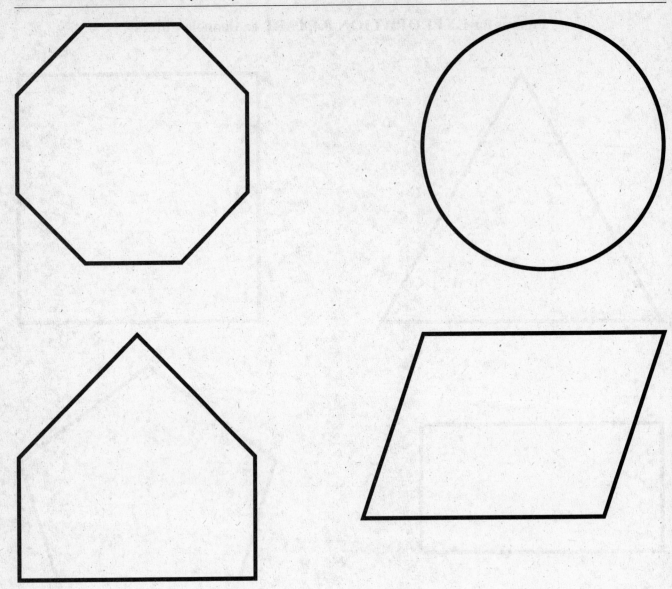

Mathematics for Elementary School Teachers
p. 470

EXPLORATION 8.5   Proof

One of the goals of the course is to emphasize the *whys* of mathematics, not just the whats (facts) and hows (procedures). The original 1989 NCTM Standards contained one standard called *Reasoning*. In the 2000 PSSM version, the standard was slightly changed to *Reasoning and Proof*. This change is significant, for it indicates the belief that proof does not begin in high school mathematics or even in middle school mathematics, but in elementary school. Earlier in the book, you have seen examples of children doing proofs: the child using Cuisenaire rods to prove that an odd number plus an odd number will always be an even number; the child proving that if you throw four darts and the numbers are all odd, then the sum must be even.

The following problem has appeared and been discussed in "Cutting Corners," "Navigating Through Geometry in Prekindergarten–Grade 2," Carole Greenes (Ed.), Reston, VA: NCTM, pp. 22–26, and in "Proof and the Middle School Mathematics Student," by James Sconyers, *Mathematics Teaching in the Middle School*, November–December, 1995, pp. 516–518. I state it this way:

1. Take a triangle, any triangle, and draw inside the triangle a line segment that connects any two points on the triangle. You now have formed two polygons. Count the total number of sides in the two polygons just formed. Do this in as many different ways as you can.

Two triangles formed
6 sides

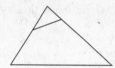

Triangle and
quadrilateral formed
7 sides

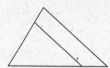

Triangle and
parallelogram formed
7 sides

2. Now do this for quadrilaterals. Three of many possible examples are given.

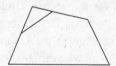

3. Repeat this process with pentagons, hexagons, and other polygons.

4. What relationship do you see between the number of sides in the original polygon and the number of sides in the two polygons just formed?

5. Prove your hypothesis.

*Mathematics
for Elementary
School Teachers*
p. 475

EXPLORATION 8.6   Using Geometric Knowledge to Make Angles

PART 1: Using Pattern Blocks

In this exploration we will simulate the development of a mathematical system using only a few "facts." Let us assume that the angles in the green Pattern Block (the equilateral triangle) are all 60 degrees. Using only this knowledge, determine the measures of each of the angles in each of the Pattern Block pieces. Write down the angles for each piece and write your "proof"—that is, your justification—for each angle.

PART 2: Paper folding

You know that the corner of a piece of paper is 90 degrees. How many different angles can you make by merely folding the paper? Write down the angles you made. For each angle, write a brief description of how you made that angle.

PART 3: Mirrors

For this exploration you will need a hinged mirror. Draw a line segment on a blank piece of paper. Using the knowledge that a complete turn is 360 degrees, how many angles can you determine?

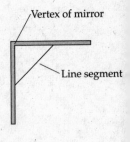

Vertex of mirror

Line segment

For example, arrange the mirror as shown in the diagram until you get a perfect square. Holding down on the mirror, draw the two line segments that form the edges of the mirror. In the mirror, you can now see four congruent squares. We know that a full turn is equivalent to 360 degrees. Therefore, by deduction, the measure of the angle you drew should be 90, because 4 congruent angles meet at the vertex of the mirror. In this manner, by creating a polygon with different numbers of sides, you can make angles whose measure you can deduce.

1.   For each angle, write the predicted measure, and then measure and write the actual angle.

2.   How close did you get to the predicted angle?

SECTION 8.2   Exploring Two-Dimensional Figures

Understanding shapes means not only memorizing definitions but also *understanding* those definitions; it means understanding relationships between shapes; it means understanding all the attributes of a shape. These explorations open up a whole new world of connections among shapes that many students find very exciting.

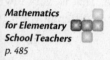

Mathematics for Elementary School Teachers
p. 485

EXPLORATION 8.7   "What Do You See?" and "Make It from Memory"

One of the things I like most about geometry is that not everyone sees the same thing.

Several of the shapes used in this exploration are found in "Image Maker: Developing Spatial Sense," in *Teaching Children Mathematics*, February 1999, pp. 374–377.[1] You will enjoy reading what the children saw!

PART 1:   What do you see?

1. Look at the shapes on page 211 selected by your instructor. For each shape do the following:
   a. Write down your response to "What do you see?" using words and phrases as opposed to full sentences.
   b. Listen to other responses.
   c. What did you learn from this experience?

2. Look at the figure at the right.
   a. What do you see? Jot down words and phrases.
   b. Find a triangle; find another one that is not congruent to the first one.
   c. Find four congruent triangles.
   d. Find four more.
   e. Find a parallelogram.
   f. Find a trapezoid.
   g. Find a pentagon.
   h. Find a hexagon.
   i. Reproduce this figure from memory.

3. Look at the second figure.
   a. How is this figure similar to and different from the one above?
   b. Find a triangle; find another one that is not congruent to the first one.
   c. Find four congruent triangles.
   d. Find four more.
   e. Find a parallelogram.
   f. Find a trapezoid.
   g. Find a pentagon.
   h. Find a hexagon.
   i. Reproduce this figure from memory.

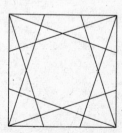

PART 2: Make it from memory

Your instructor will show you various shapes, with a catch.

You will get to see each shape for only a few seconds before the instructor covers it. That is, you will have to draw it from memory.

In each case do the following:

1.   Sketch the shape.

2.   Make any changes as you see the shape for the second and third time.

3.   Think of what went on in your head that enabled you to make the shape. That is, very few people see the whole shape. They break it apart and put it back together in different ways.

4.   Listen to other student responses.

5.   What did you learn from this exploration?

PART 3: Giving directions

Your instructor will select a pattern and then ask you to write directions for making that pattern so that someone hearing the directions on the phone could accurately reproduce the figure (don't worry about size).

What do you see?

Mathematics for Elementary School Teachers
p. 487

EXPLORATION 8.8 Making Shapes from Folding a Square

A key goal related to developing geometric understanding is to see relationships among shapes. Toward this end, in Tape 8 in the Annenberg Teaching Math K–4 video series, a second-grade teacher gives his students a square piece of paper that they fold three times. He then asks them how many different shapes they can make with this piece of paper. The only rule is that they can use only the folds that they have made—that is, no new folds. There is a lot of mathematics in this exploration. If you have access to this video, it is very interesting viewing.

The exploration you will do is more sophisticated but similar.

1. Take a blank sheet of paper that measures $8\frac{1}{2}$ by 11 inches. How can you determine how much to cut off—without using a ruler—so that you will have a square piece of paper? Do it.

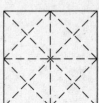

2. Now that you have a square sheet of paper, make the folds illustrated at the right.

3. Determine how many different shapes you can make, using only the folds given. Several of the NCTM process standards are illustrated in this exploration.

First, one goal is to get as many shapes as possible (Problem Solving).

Second, you will get more shapes if you have some kind of systematic approach, at some point, as opposed to just random folding (Reasoning).

Third, you will get more shapes if you think outside the box; recall the 9-dots investigation in Chapter 1 (Reasoning).

Fourth, you need to think about how you will display your results (Representation/Communication). If you simply sketch them roughly, you might miss some. For example, there are several squares that can be made that are different sizes. Also, when you compare your answers with other members of the class, some thought about how you want to represent your results will make it easier to find answers you have in common, and answers you have that your neighbor doesn't have, and vice versa.

a. After you are finished, summarize your approach. This should be a short paragraph—enough detail so that a reader who only got a few shapes could understand your strategies and be able to get many more shapes, but not a blow-by-blow account of your thinking for each shape.

b. Explain why you chose the representation system that you selected to display all the shapes you found.

4. Summarize what you learned from this exploration.

a. What mathematics did you learn?

b. If you learned new ideas about problem solving, describe them.

c. If you learned new ideas about reasoning, especially thinking systematically, describe them.

d. If you learned new ideas about communication, describe them.

e. If you learned new ideas about representation, describe them.

f. If you see connections that you hadn't seen before, describe them.

g. If you find that your attitudes and/or beliefs about geometry have changed, describe the change.

EXPLORATION 8.9   Definitions and Language

This exploration is designed to help you see definitions from a different perspective. Follow this process for each term you are asked to define. Additional steps are given for specific terms.

Step 1:   Write your "first draft" definition, in your own words, of this term.

Step 2:   With your partner(s):
  a. One person reads his or her definition.
  b. The other persons first comment on validity; that is, do they think the definition is valid? If not, discuss the definition until all members agree that the definition is valid or invalid. If all agree that it is invalid, then discuss how to change it to make it valid.
  c. The other persons now comment on clarity; that is, are there aspects of the definition that are not clear or are ambiguous? Discuss those aspects of the person's definition. Work together to rewrite words or phrases that are ambiguous or not clear.
  d. Move to the second person and repeat the process.

Step 3:   Class discussion

Step 4:   What did you learn about this term and/or what did you learn about the process of defining terms?

1.   Do Steps 1–4 for the term *perpendicular*.

2.   Do Steps 1–4 for the term *angle*.

   Now read the article "The Role of Definition" in *Mathematics Teaching in the Middle School*, April 2000, pp. 506–511. See the many different ways in which the children thought about angle. What did you learn from reading the article?

3.   Do Steps 1–4 for the term *adjacent angle*.

   Now look at the definition of this term in the textbook. Does it make sense? If it was different from yours, why do you think I (the author) chose that definition?

4.   Do Steps 1–4 for the term *diagonal*.

5.   Do Steps 1–4 for the term *quadrilateral*.

6.   Do Steps 1–4 for the term *polygon*.

Step 5:   What is wrong with the following definition: A polygon is a geometric figure with a certain number of sides. Can you make a counterexample that fits the definition but is not a polygon?

Step 6:   Here are several other definitions. Discuss the definitions. What do they all have in common? Which one do you like best? Why?
   Polygon: the union of several line segments that are joined together so as to completely enclose an area.
   Polygon: a closed plane figure with *n* sides.
   Polygon: plane closed figure whose sides are straight lines.

7. Below are several examples of concave and convex polygons. Look and see what it is that all convex figures have in common and what it is that all concave figures have in common. I will tell you ahead of time that there is not one "right" way. Now do Steps 1–4 for the term *convex*.

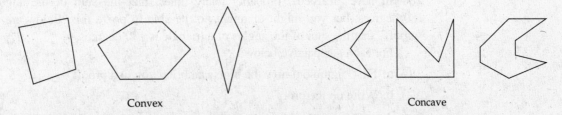

    Convex   Concave

    Step 5: Give your definition to a roommate or friend who is not taking this course. Ask him or her to tell you what *convex* and *concave* mean. Ask for feedback.

    Step 6: What if we didn't limit ourselves to polygons but included simple, closed curves where not all the sides were line segments (for example, curves, ellipses, and figures like the one at the right). Would all of the definitions of convex apply to non-polygons or would some apply better than others?

8. Consider the term *corner*. How many corners does a cube have? Defining *corner* was not originally in this exploration because mathematical dictionaries don't define *corner*. Then I read an article called "What Makes a Corner a Corner?" in *Teaching Children Mathematics*, September 1998, pp. 6–9. The teacher didn't realize that *corner* is not a mathematically defined word, and then she realized that her students were becoming confused. However, she stayed with their thinking and some neat things came from the quest.

    Do Steps 1–4 for the term *corner*.

    Step 5: How is *corner* related to the mathematical terms *vertex*, *edge*, and *face*?

    Step 6: Read the article. What did you learn from reading the article?

9. Do Steps 1–4 for the term *cylinder*.

*Mathematics
for Elementary
School Teachers*
p. 495

EXPLORATION 8.10   The Sum of the Angles in a Polygon

PART 1: Triangles

You all have "learned," probably many times, that the sum of the angles of a triangle is 180 degrees, but you might or might not be able to prove this. Below are sketches of several "proofs" that the sum of the angles of a triangle is 180 degrees.

For each perspective below:

a. First, figure out how the hints can bring you to a proof.

b. Write up the proof.

1.   A favorite approach in elementary school is to label the angles and tear off the edges. Then rearrange the three angles in a creative way to show that their sum is 180 degrees.

2.   A classic proof is begun below:

Begin with triangle *CAT*.

Draw a line through *T* parallel to *CA*.

Label the angles.

Knowing relationships with angles that are formed with parallel lines, we can now prove that $m \angle 1 + m \angle 2 + m \angle 3 = 180$ degrees.

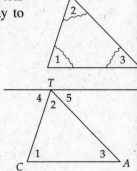

3.   Another classic proof is sketched below:

Begin with a triangle. Fold the triangle so that the top vertex lies on the base and so that the fold line is parallel to the base. A number of new angles are formed. Look for relationships between the original angles and the new angles.

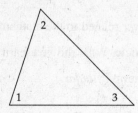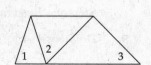

PART 2: Quadrilaterals

Prove that the sum of the angles of any quadrilateral is 360 degrees.

PART 3: Polygons

Prove that the sum of the interior angles of any polygon is equal to $180(n - 2)$, where *n* represents the number of sides in the polygon.

Below is an illustration of this theorem that is not equivalent to a proof because it simply accepts that, but does not justify why, the sum of the angles increases by 180 each time you add a side.

Triangle	Sum = 180	180×1
Quadrilateral	Sum = 360	180×2
Pentagon	Sum = 540	180×3
Hexagon	Sum = 720	180×4
n-Gon		$180 \times (n - 2)$ because in each case, the number you multiply 180 by is 2 less than the number of sides of the polygon.

EXPLORATION 8.11 Congruence

The notion of congruence is one of the "big ideas" in mathematics and has important applications in industry, science, and art. The whole idea of mass production involves making many congruent copies. When you have a mold, you have more assurance that all copies are congruent. Before mass production, this was an issue. Even with mass production, engineers work with "tolerances." How much can something be "off" and still work. For example, tolerances for pistons in cars are measured to the ten-thousandth of an inch!

In this exploration we will explore, with different manipulatives, the notion of how we determine figures to be congruent at higher van Hiele levels than "it fits."

PART 1: Geoboards

1. **a.** Divide this region into 4 congruent triangles.
    b. Divide this region into 3 congruent triangles.
    c. Divide this region into 3 shapes that have the same area.

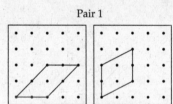
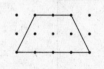

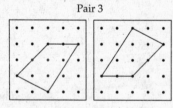

2. Divide your whole Geoboard into two congruent regions. Represent your answers on Geoboard Dot Paper (in the back of the book).

    Do it again, a different way. Do it again, a different way. Do it again, a different way. How many different ways do you think there are?

    Once your class has a number of examples, describe what all the figures have in common, besides being half. Look at the shapes.

3. Look at the following pairs of figures. Are they congruent or not? Justify your answer.

Pair 1   Pair 2   Pair 3

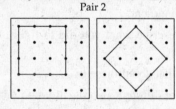

    a. Which, if any, pairs are congruent?

    b. How can you tell without cutting them out and placing one figure on top of the other?

    c. Discuss the question asked in Step b, and then develop a test that someone else could use to determine whether two figures are congruent.

PART 2: Tangrams

1. We can make a trapezoid with the tangram's parallelogram and the two little triangles, and we can make a trapezoid with the tangram's medium triangle and the two little triangles. Are the two trapezoids congruent? How could you prove that they are or are not?

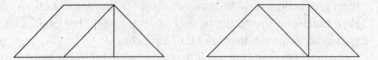

2. Are the members of each of these pairs of tangram figures congruent? Explain your reasoning.

   a.

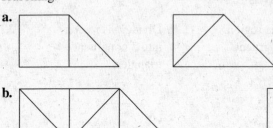

   b.

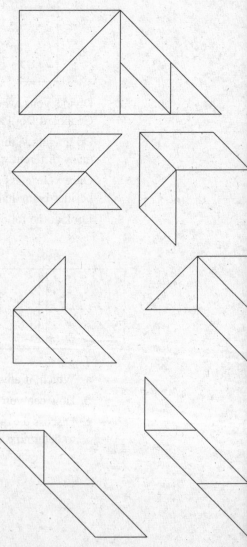

3. Here are two shapes from Exploration 8.2, Part 4, where you were asked to make as many tangrams as you could from the parallelogram and the two little triangles.

   a. Without cutting them out, determine whether the two figures below are congruent.

   b. Describe what you did to figure this out.

   c. What about these two?

4. Three different words are commonly used by students when describing figures that have a certain amount of sameness: *same*, *congruent*, and *identical*. Some people argue that having three different words is picky. Recall that in Chapter 3 we found that many teachers prefer not to use *borrowing* and *carrying* but rather use one word for both operations. What do you think? Do we need all three words? Why or why not?

EXPLORATION 8.12   Polygons with Various Attributes

Your instructor will either specify the tools you can use or allow you to choose the tools that you will use: thin strips of file folder paper that you can cut and tape together, ruler, protractor, compass, and/or software.

In Section 8.1, you read that two points *determine* a line. With respect to polygons, we will also speak of conditions that *determine* a specific polygon. For example, if you ask how many different kinds of quadrilaterals you can make that have four right angles, there are two that fit this condition: rectangles and squares. Thus, having four right angles does not *determine* a square; we need to say more.

In each case, predict what you think you will get and briefly explain your prediction. After you have finished:

If you believe that there is exactly one answer, explain why.

If you believe this is impossible, explain why.

If you believe that there is more than one answer, explain why.

1. *Triangles*

   Determine how many different triangles can be made with

   a. these three lengths: 2 cm, 3 cm, 4 cm.

   b. these three lengths: 3 cm, 5 cm, 9 cm.

   c. one angle 40 degrees, one angle 60 degrees, and one angle 80 degrees.

   d. two sides 4 cm and a 90-degree angle.

   e. one side 3 cm, one side 4 cm, and the angle between them 45 degrees.

   f. one side 3 cm, one side 4 cm, and one angle (not the one between the two specified sides) 45 degrees.

2. *Quadrilaterals*

   How many different quadrilaterals can you draw that have

   a. these four lengths: 2 cm, 3 cm, 4 cm, 5 cm?

   b. all sides equal?

   c. all angles equal?

   d. all sides equal but not all angles equal?

   e. all angles equal but not all sides equal?

   f. exactly 2 pairs of congruent sides, but each pair having different length?

   g. exactly 2 congruent sides?

   h. exactly 3 congruent sides?

   i. exactly 1 right angle?

   j. exactly 2 right angles?

   k. exactly 3 right angles?

3.   *Pentagons*

How many different pentagons can you draw that have

   a. all sides equal?

   b. all angles equal?

   c. all sides equal but not all angles equal?

   d. all angles equal but not all sides equal?

   e. exactly 3 congruent sides?

   f. exactly 4 congruent sides?

   g. exactly 2 right angles?

   h. exactly 3 right angles?

   i. exactly 4 right angles?

4.   *Hexagons*

How many different hexagons can you draw that have

   a. all sides equal?

   b. all angles equal?

   c. all sides equal but not all angles equal?

   d. all angles equal but not all sides equal?

   e. exactly 3 congruent sides?

   f. exactly 4 congruent sides?

   g. exactly 5 congruent sides?

   h. exactly 2 right angles?

   i. exactly 3 right angles?

   j. exactly 4 right angles?

Mathematics for Elementary School Teachers pp. 494, 502

EXPLORATION 8.13 **Polygons and Relationships**

1. Cut out the shapes on page 223. This set of quadrilaterals contains many different kinds of quadrilaterals, such as a "skinny" rectangle and a rectangle that is "closer to" a square.

   a. Pull two shapes at random and describe how they are similar (attributes they have in common) and how they are different (attributes they don't have in common). Compare your responses with those of your partner(s) or another group. Repeat this as many times as your instructor indicates.

   b. Pick a shape. Don't let anyone see it. Your partner(s) need to guess which shape you have by asking only questions that can be answered yes or no. How long does it take to guess which shape you picked out? Repeat this as many times as your instructor indicates.

   c. What guesses seem to be better first guesses than others?

   d. Let's say you were guessing and you had the shape narrowed down to square, rhombus, or rectangle. What would be your next question? Why? Could you get it with one question?

2. a. Make a table for the shapes. The first column will contain the letter denoting each shape. The other columns will consist of attributes that some, but not all, of the quadrilaterals have (such as parallel sides) or will contain an attribute that quadrilaterals have in different numbers (such as the number of right angles).

   b. Your instructor will either have the class determine the columns first or have each group determine the columns and then present their findings to the class.

   c. Summarize what you learned from this step.

3. *Venn diagrams*

   Below are several Venn diagrams. In each case, first describe the characteristics of all the quadrilaterals that are in the same ellipse or circle. Then write the missing labels. In some cases, this will be a word; in other cases, it will be a phrase. Note that these are adapted from the NCTM book *Navigating Through Geometry in Grades 3–5!*

   a.
   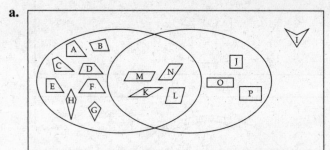

   b.
   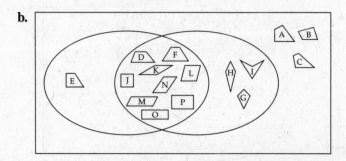

c.

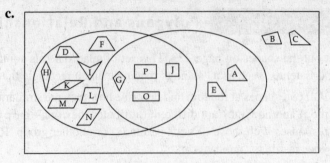

d.

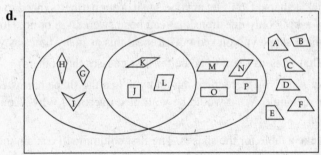

e.

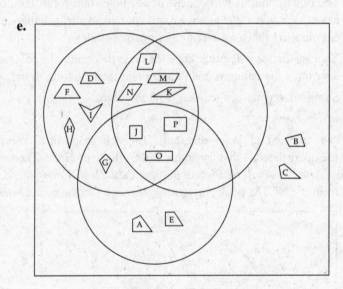

f.

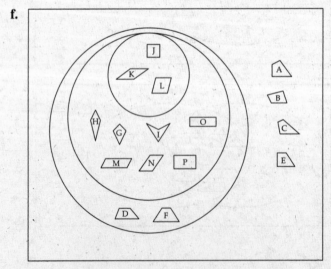

4.   Make up your own problems.

Figures for EXPLORATION 8.13: Polygons and Relationships

SECTION 8.3   Exploring Three-Dimensional Figures

When we make complex objects, whether they be cars or houses or sculptures, they have to be designed first, and the actual construction is done from blueprints or drawings—that is, two-dimensional representations of the object. More and more textbooks and other curriculum materials are emphasizing spatial visualization. It is very likely that you will do explorations like the ones below, at a simpler level, with your future students. The following explorations provide you an opportunity to work on these concepts and ideas and to do so in a way that should be both enjoyable and challenging.

Mathematics for Elementary School Teachers
pp. 520, 521

EXPLORATION 8.14   Exploring Polyhedra

PART 1: Comparing polyhedra and polygons

Your instructor will direct you to a set of polyhedra—either physical models or pictures.

1. How are polyhedra like polygons and how are they different from polygons?

2. Focus now on the set of prisms. What attributes do they *all* have in common?

3. Focus now on the set of pyramids. What attributes do they *all* have in common?

4. How are prisms like pyramids and how are they different from pyramids? You might use a Venn diagram to represent your ideas!

5. In this step, you explore what terms used to describe polygons are also used to describe polyhedra, what new terms are needed, and why.

    a. If a polygon is a simple, closed curve, is there an analogous definition for polyhedra?

    b. All polygons have vertices, line segments, and angles. Do these terms work for describing polyhedra? Do we need new terms too? If so, why?

    c. With some polygons, we use the terms *base* and *height*. Do those terms apply to all polyhedra?

    d. We can call a polygon with n sides an n-gon. Is there an analogous notion for polyhedra (i.e, is a polyhedron with n sides an n-dron)? Why or why not?

    e. If there are regular polyhedra, how might they be defined?

6. When we want to describe one of the line segments of a polygon, we use the term *side*. However, when describing one of the line segments of a polyhedron, mathematicians use the term *edge*. Why do you think this term is used instead of *side*?

7. When describing polyhedra, many people use the term *corner*, yet this term is not found in mathematical dictionaries. Why do you think that is true? Define *corner* in your own words.

8. a. We speak of convex and concave polygons. Do those terms apply to polyhedra? Support your thinking.

    b. If you defined *convex* in Exploration 8.9, look at the definitions that were created. Do some of the definitions apply to three-dimensional figures better than others do?

PART 2: Determining congruence with polyhedra

You can read childrens' solutions to this problem in the April 2002 issue of *Teaching Children Mathematics*, pp. 444–445.

1. Each group will make at least two different polyhedra from two squares and six equilateral triangles, where the lengths of the sides of the squares and of the equilateral triangles are equal.

2. Describe your thinking process for making the polyhedra and in going about making another one that was different from the first one you made.

3. Examine the class set. How many different polyhedra are there in the set? Here *different* means "not congruent."

    If two structures were close to congruent, what did you do to determine whether they were in fact congruent or not congruent?

4. Determine the number of edges, vertices, and faces for the different figures. You will use this information in Exploration 8.15.

PART 3: Sort polyhedra into groups

1. Take your set of polyhedra or the set given by your instructor.

2. Sort the polyhedra into two or more groups. Describe the criteria for each group. If your instructor gives you a new object, determine which of your groups it goes into. Discuss this until all students agree or until there is an impasse.

3. Repeat the process as many times as your instructor indicates.

PART 4: Writing directions

1. Write directions for making a structure that you have selected or a structure selected by your instructor.

2. Compare descriptions.

3. a. Give the directions to someone who is not in this class. Have this person try to make the polyhedron from your directions.

    b. If problems arose, describe and analyze the problem areas. What caused the confusion?

    c. Revise your directions, if necessary.

Mathematics
for Elementary
School Teachers
pp. 523, 527

EXPLORATION 8.15 Relationships Among Polyhedra

Leonard Euler discovered a relationship among the number of vertices, the number of edges, and the number of faces of any *polyhedron*. Rather than present this relationship, I ask you to look at a variety of polyhedra.

Collecting data and making hypotheses

1. Count and record the numbers of vertices, edges, and faces in each polyhedron shown below. You can use pictures of other polyhedra in the text if you wish. You can also use the work from Exploration 8.14, Part 2, if you did that.

2. What patterns do you see, and what observations or hypotheses do you have? Record them in the table.

3. Describe the relationship you found among the numbers of vertices, faces, and edges of any polyhedron.

4. Describe any significant actions you took that facilitated this discovery.

a.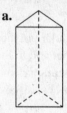

b.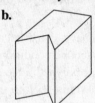

c.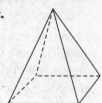

d.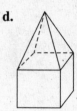

e.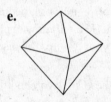

f.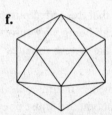

g.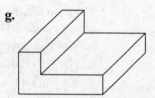

*Mathematics
for Elementary
School Teachers*
p. 526

EXPLORATION 8.16   Regular Polyhedra

1. In Section 8.2, we defined a regular polygon as a polygon in which all sides have the same length and all interior angles have the same measure. As we have seen, some definitions with two-dimensional figures extend nicely to three-dimensional figures and some do not. Can we directly extend the definition of *regular polygon* to *regular polyhedron*? That is, can we define a regular polyhedron as a polyhedron in which all the sides are congruent and all the interior angles are congruent? What do you think?

2. The Greeks chose to define a regular polyhedron as a figure in which all the faces are congruent regular polygons and the numbers of edges meeting at each vertex are the *same*.

   a. How are this definition and the one proposed in question 1 similar? How are they different?

   b. Why do you think the second condition of the regular polyhedron, "same number of edges meeting at a vertex," was chosen instead of the second condition for the regular polygon, "same number of degrees at each vertex"?

3. How many regular polyhedra do you think there are? Briefly describe the thinking behind your guess.

4. Using the materials you have, construct as many regular polyhedra as you can. Stop at some points to make sure the second condition is being met and to look at the development of the figure—all of the regular polyhedra have lots of symmetry, and you can see this symmetry develop as you make the figures! Note any observations, discoveries, and questions to bring up at the class discussion.

5. You have discovered that the actual number of regular polyhedra is much smaller than most of you guessed. (I recall being very surprised!) We will now explore why this is so. But first, before reading on, take a few minutes in your groups to discuss why there are so few regular polyhedra.

6. As we build the polyhedra, it turns out that there are two conditions that *all* convex polyhedra, not just regular polyhedra, must obey. The first condition is that at any vertex, the sum of the measures of the angles must be less than 360 degrees. The second is that there must be at least three polygons at any vertex. Take some time in your small group to discuss why these conditions must be true. Record your thinking about

   a. why the first condition must be true.

   b. why the second condition must be true.

7. This step will help to advance your thinking.

   a. If you were to make a regular polyhedron with equilateral triangles, what are the possibilities for the number of triangles at each vertex? Explain each case.

   b. If you were to make a regular polyhedron with squares, what are the possibilities for the number of squares at each vertex? Explain each case.

   c. If you were to make a regular polyhedron with regular pentagons, what are the possibilities for the number of pentagons at each vertex? Explain each case.

   d. If you were to make a regular polyhedron with regular hexagons, what are the possibilities for the number of hexagons at each vertex? Explain each case.

EXPLORATION 8.17 Block Buildings

Here we will explore different possible ways to represent a three-dimensional object on paper. We will begin with simpler objects—objects that can be made from stacking congruent cubes—and then move to increasingly complex figures.

PART 1: How many cubes?

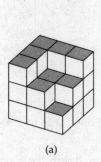 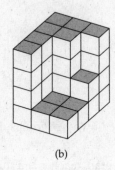 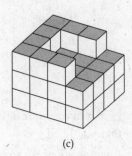 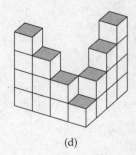

(a)   (b)   (c)   (d)

1. **a.** Look at structure (a). How many cubes were needed to make the structure? Work on this alone and briefly note your work.

   b. Compare strategies with other members of your group. Note any strategies that you like.

2. **a.** Look at structure (b). How many cubes were needed to make the structure? Work on this alone and briefly note your work.

   b. Compare strategies with other members of your group. Note any strategies that you like.

3. **a.** Look at structure (c). How many cubes were needed to make the structure? Work on this alone and briefly note your work.

   b. Compare strategies with other members of your group. Note any strategies that you like.

4. Write the minimum and maximum number of cubes that could be contained in structure (d) above, if it is viewed from "ground level." Justify your answer.

PART 2: Making directions

A very major role that mathematics has played in the development of much technology is a process that is called *mathematizing*—that is, taking a real-world situation or problem and using mathematics to understand that situation or problem better. If we make a block building and other people want to make a copy of that building, they can come to us and make the copy right next to ours. If that is inconvenient, we could take a photo and they could make a copy from the photo. However, as you just found out, photos can't show everything. Of course, we could photograph the building from all sides. That would solve the block problem, but not, for example, blueprints for a house, topographical maps, and other things.

1. **a.** Look at the block buildings on this page and the next page. Think of a way to communicate how to construct that building to another group. Your way may consist of words and/or diagrams.

   b. Give your directions to another group and have them construct the building from your directions. Note any parts that they have difficulty with—that is, that they find confusing and/or ambiguous.

c. Switch roles so that your group constructs a building from their directions. Note any parts that you have difficulty with—that is, that you find confusing and/or ambiguous.

d. Repeat this process until you have developed a method that will enable another group to construct an exact copy of your block building successfully.

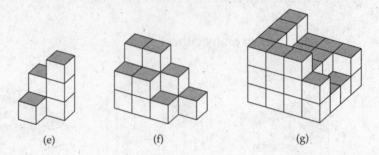

(e)   (f)   (g)

2. Listen to methods developed by other groups. Note any ideas or comments that you heard that you liked.

3. Imagine you were an elementary teacher and students all over the country were making such buildings and comparing their work with other students. If you could choose among the different methods (via fax), which would you choose? Your response needs two parts:

a. An analysis of the pros and cons of each method

b. Your choice and a justification of your choice

PART 3: Making drawings with isometric dot paper

We have explored different ways to communicate how to make block buildings. In this exploration, we will explore how to draw them. A special kind of dot paper (see Isometric Dot Paper at the end of this book) has been created to make such drawing easier. Make several copies of that page.

1. Draw the figure at the right on your dot paper. Many students find it takes several rounds of guess–check–revise before their picture matches the one at the right. Once you succeed,

a. describe strategies that you discovered while copying the shape.

b. note other observations, insights, and questions (including what-if questions).

2. Practice with the figures below and/or make block buildings and then draw them. After these further explorations,

a. describe new strategies that you discovered while making your drawing.

b. note other observations, insights, and questions (including what-if questions).

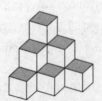
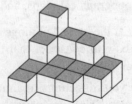
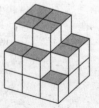

3.   Make a picture of a building that contains between 10 and 15 cubes and will be "challenging" to draw, for example, a building with fewer cubes in a back row than in a front row. Write the directions for making that building. Explain any difficulties and how you overcame them.

PART 4: Continuation

There is another method for giving directions for making block buildings. Rarely do students come up with this method in Part 2, and yet it is a method that is commonly presented to children in elementary and middle schools. One reason for this is that although it is not an obvious way, there is much potential for learning by examining this method. This method comes from the idea of taking a picture from all sides. For example, consider the building below.

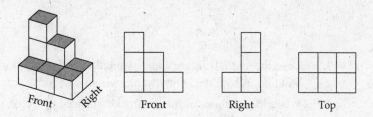

To the right of the building are three views of the building: the front view, the right side view, and the top view. To understand this method, some students find it more useful to imagine looking at the building from ground level. Envision crouching down so your eye is at the same level as the bottom of the building. Imagine seeing only the shadow of the building, like seeing someone's silhouette.

1.   Make the buildings from the three views given in the exercises below. Afterwards, you will be asked to evaluate this method by comparing it to the methods you developed earlier.

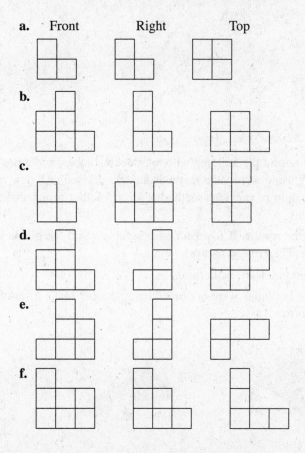

2.   Draw the front, right side, and top views of the buildings below.

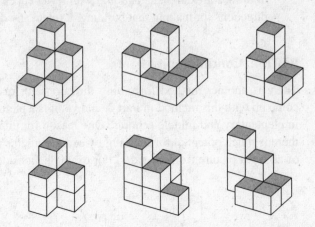

3.   Describe the advantages and strengths of this method of giving directions for making a building. What kinds of buildings would it be most suited for, or is it an all-purpose method?

4.   What do you think of this method compared to the ones you developed earlier?

PART 5: What if

1.   What if you could make buildings with interlocking cubes—for example, the one at the right?

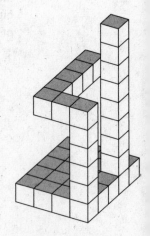

   a. Make a nonsimple building with interlocking cubes, and write directions for making the building using any of the methods already devised or a new method (which could be a modification of previous methods). Have another group make the building using your directions.

   b. Report the results. If it wasn't successful, explain the glitches and how you solved the glitches. Then repeat part (a).

   c. What did you learn from this?

2.   What if the buildings were not restricted to cubes? Make a set of directions for one of the buildings below.

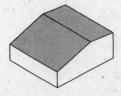  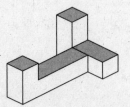

Mathematics
for Elementary
School Teachers
p. 530

EXPLORATION 8.18   Cross Sections

One important spatial visualization skill is being able to imagine what happens when physical objects are changed or moved. Examples include being able to imagine how a new arrangement of furniture will work in a living room and being able to figure out whether all of a set of objects will fit into a given space (such as a suitcase or the trunk of a car).

Here, we will explore the notion of cross sections. Consider the cube at the right. If we sliced it in half (as though a knife made a vertical slice), what would the newly exposed face (that is, the cross section) look like? If we sliced it at an angle, what would the new face look like now?

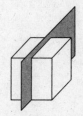

PART 1: Predictions

For each figure below, first predict what the sliced face will look like. Then check your prediction either by making a clay figure and slicing it or by some other means. Describe any observations.

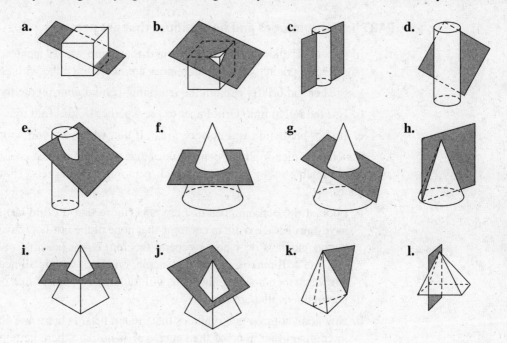

PART 2: Descriptions of intersections

In Part 1, you were given a picture representing a specific cross section of a solid object. What if you had no picture?

1.   Describe an object and an intersecting plane, using only words.

2.   Have another group describe what slice they would make on the basis of your description. If they are able to decode your directions accurately, great. If there are problems, analyze them, and then revise the description to address the problems raised.

*Mathematics
for Elementary
School Teachers*
p. 531

EXPLORATION 8.19   Nets

This exploration continues our work in looking at connections between two-dimensional and three-dimensional objects. A net is simply a two-dimensional figure that will fold up into a three-dimensional figure. For example, the shape below at the left is a net for a cube, because it will fold up into a cube. The shape at the right below is not a net, because it will not fold into a cube or other three-dimensional figure.

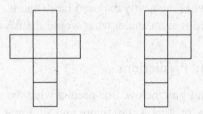

PART 1: Pentominoes and hexominoes that are nets

1.   If you did Exploration 8.3, you found that there are 12 pentominoes. Use the table on page 235 to predict which pentominoes are nets for a cube with no top.

   a. Predict and briefly explain the reasoning behind your prediction.

   b. Use the Polyomino Grid Paper to see whether it does fold up.

   c. If your prediction was correct, great. If not, why were you wrong? What did you learn?

2.   It would be nice if all the pentominoes that are nets had one characteristic in common that would enable us to remember which pentominoes are nets. However, this is not always the case.

   a. Look at the pentominoes that are nets (those that do fold up) and see whether they all have one characteristic in common that none of the not-nets have. If not, you will do what mathematicians do when a general theorem is not possible or has not been discovered yet: You will make cases. For example, one such case is all nets that have four cubes in a row and then one on either side will fold up. Do you see that one? What about the other pentominoes that are nets?

   b. Now look at those pentominoes that are not nets. Is there one characteristic they all have in common that none of the nets have? If not, are there certain characteristics that preclude the possibility of folding into a cube?

3.   If you did Exploration 8.3, you found that there are 35 hexominoes. Look at the table on page 237 showing all 35. Using what you learned about nets, name those hexominoes that you are close to 100% sure will fold up into a cube. Explain why. Then name those hexominoes that you are not close to 100% sure will not fold up into a cube. Explain why not.

4.   Now look at the hexominoes that do fold up into a cube. Once again, is there one characteristic they all have in common that none of the not-nets have? If not, are there certain characteristics that some of the nets have in common, as we found for pentominoes?

Pentominoes That Fold into Cubes

Name	Prediction and reasoning	Yes/No	Reflection
F			
I			
L			
N			
P			
T			
U			
V			
W			
X			
Y			
Z			

Hexomino Sheet

PART 2: Making nets

One of the most common nets that people (especially those who recycle) experience regularly is a flattened cereal box.

1. Here is one set for a standard cereal box.

   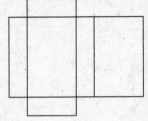

   a. What do you notice about this net?

   b. What attributes and characteristics does it have?

2. Now make as many *different* nets as you can for the cereal box. If you use Polyomino Grid Paper, you can make the front and back 4 × 2 rectangles, the sides 4 × 1 rectangles and the top and bottom 2 × 1 rectangles.

3. Examine the whole class set of nets for a cereal box. What do you see? Are some of these more similar than others—that is, in the same family, just as a triangular prism, a cube, and a box are all in the same family that we call "prism"?

4. What attributes do all of the nets have in common?

5. Do you see any connections among the dimensions of each side?

6. Do you see any connections between the nets for cereal boxes and our work with pentominoes and hexominoes? If so, describe the connections.

7. Make as many nets as your instructor asks for the following shapes.

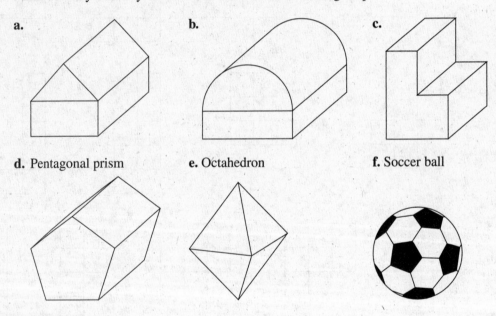

a.

b.

c.

d. Pentagonal prism   e. Octahedron   f. Soccer ball

PART 3: Will the net fold up?

Now that you have experience making nets for shapes, let's go the other way. Following are some possible nets. Predict which will fold up into a polyhedron and which will not. Give your reasons. If you predict it will, sketch the polyhedron or describe it.

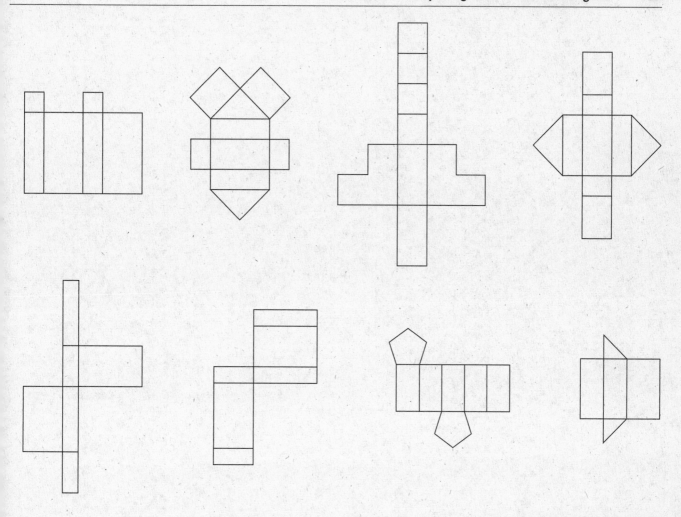

PART 4: Making a package for an object

We see packages everywhere. Most packages are manufactured as flat surfaces that are then cut and folded and glued or stapled to become packages.

1.   Select an object for which you will make the package. It might be similar to a package you have seen, and you want to improve or modify it, such as a package for french fries. It might be something not normally packaged, such as a bunch of bananas. You can consider large objects, in which case you would make a scale model package.

2.   Think about the package—the shape, how much space you want, padding, whether the object should be seen, how the package will be opened and (if necessary) closed again.

3.   Make the package. Turn in

     a. the package

     b. the net for your package

     c. your explanation of why this package is "right" for this product—a sentence or two

4.   Describe how you used your spatial sense, estimating skills, and measuring skills in your design—a paragraph that essentially describes what mathematical knowledge and skills you used in this project.

CHAPTER 9

Geometry as Transforming Shapes

A s in Chapter 8, the first three explorations use Geoboards, tangrams, and polyominoes to explore topics from throughout the chapter.

Mathematics for Elementary School Teachers p. 545

EXPLORATION 9.1 Geoboard Explorations

PART 1: Slides, flips, and turns

1. **a.** Write directions for moving each of the figures from position A to position B.

    b. Exchange directions with someone else.

    c. On the basis of the feedback you receive, either keep your response to part (a) or write a second draft.

(i) 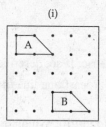 (ii) 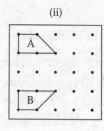 (iii) 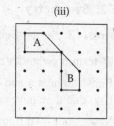 (iv)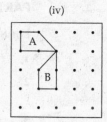

2. **a.** Write directions for moving each of the figures below from position A to position B.

    b. Exchange directions with someone else.

    c. On the basis of the feedback you receive, either keep your response to part (a) or write a second draft.

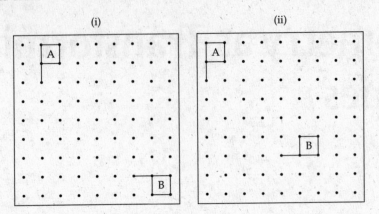

3. Use the figures on page 245. In each problem, reflect the polygon across the given line and sketch the reflection of the figure.

4. For each of the following, use a Geoboard or Geoboard Dot Paper.

    a. Make a figure that will look the same when we turn the Geoboard 90 degrees (a $\frac{1}{4}$ turn), 180 degrees (a $\frac{1}{2}$ turn), and 270 degrees (a $\frac{3}{4}$ turn).

    b. Make a figure that will look the same when we turn the Geoboard 180 degrees but different when we turn it 90 degrees.

    c. Make a figure that will look the same when we flip the Geoboard across a horizontal line (so that the figure is turned over). *Note*: In this case, you have to imagine a transparent Geoboard.

    d. Make a figure that will look the same when we flip the Geoboard horizontally but different when we flip it vertically.

    e. Make a figure that will look the same no matter how we flip or turn the Geoboard.

PART 2: Symmetry

For Steps 1–3, use the figures on page 247.

1. Determine whether each figure is symmetric. If so, describe the line(s) and/or rotation(s).

2. Finish figures a and b so that they have at least one line of symmetry. Finish figures c and d so that they have rotational symmetry.

3. Make your own figures for a classmate to test for symmetry.

4. Make a square on Geoboard Dot Paper.

    a. Modify one side. Now modify the opposite side so that the figure still has symmetry.

    b. Modify another side. Now modify the opposite side so that the figure still has symmetry.

    c. Repeat parts (a) and (b) with other squares and different modifications. What generalizations can you make?

    d. Repeat parts (a) and (b) with other symmetric polygons that have an even number of sides. What generalizations can you make?

5. Make a square on Geoboard Dot Paper.

   a. Modify the top side. Now modify the side to the right so that the figure still has symmetry. Repeat the process, making the same modification each time, until all four sides of the square have been modified.

   b. Repeat part (a) with other squares and different modifications. What generalizations can you make?

   c. Repeat this with other symmetric polygons with an even number of sides. What generalizations can you make?

PART 3: Similarity

1. Use the figures on page 249 and several sheets of Geoboard Dot Paper. For each figure, make a larger copy of the shape on a sheet of Geoboard Dot Paper so that the original and the larger shape are "similar." Briefly describe how you know that they are similar.

2. Compare your results with those of other members of your group.

3. Draw your own shapes on Geoboard Dot Paper, and then make larger copies as in Step 1.

4. Write a definition of the term *similar*. That is, develop a definition that someone else could use to determine whether two figures are similar. Another way of thinking about this question is to determine what relationship(s) a smaller and a larger figure must have in order to be similar.

Figures for EXPLORATION 9.1, PART 1: Slides, flips, and turns

3. a.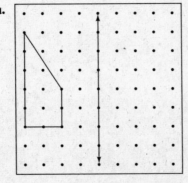

b.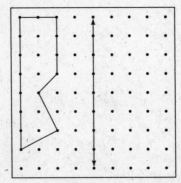

c.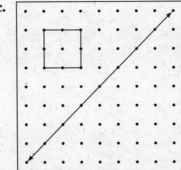

d.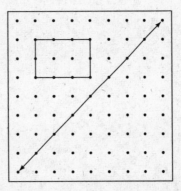

Figures for EXPLORATION 9.1, PART 2: Symmetry

1. a. 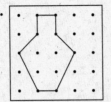   b.

    c. 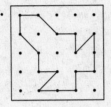   d.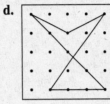

2. a. 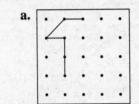   b.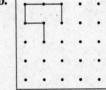

    c. 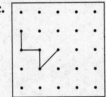   d.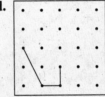

3. a. 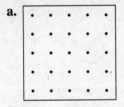   b.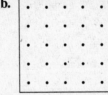

    c. 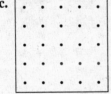   d.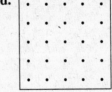

Figures for EXPLORATION 9.1, PART 3: Similarity

a.

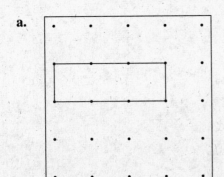

b.

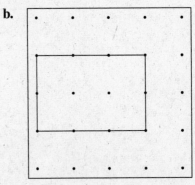

c.

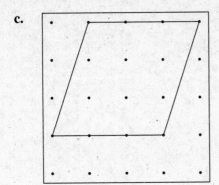

d.

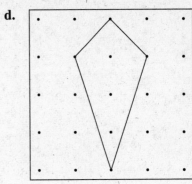

e.

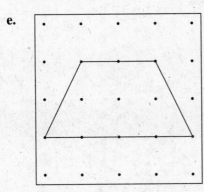

f.

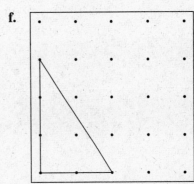

g.

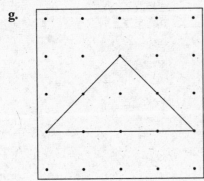

h.

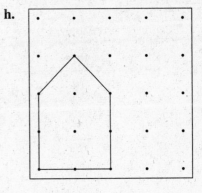

Mathematics for Elementary School Teachers
p. 545

EXPLORATION 9.2 Tangram Explorations

PART 1: Symmetry

1. Make different shapes from your tangram pieces so that each overall shape has one line of symmetry.

2. Make different shapes from your tangram pieces so that each overall shape has two lines of symmetry.

3. Make different shapes from your tangram pieces so that each overall shape can be rotated and still look the same. Specify the nature of the rotation symmetry of the shape.

PART 2: Similarity

Rectangles

1. **a.** With your group, make as many different rectangles as you can, using the pieces from one tangram set. Make a copy of each rectangle by tracing it.

   b. Compare your results with those of another group (or of the whole class). Add any new drawings of rectangles to your group's collection.

2. **a.** Separate the set of rectangles into subsets so that all the rectangles in a subset are similar to each other.

   b. Describe how you determined which rectangles are similar. Then write a definition or a rule for *similar* rectangles. This will be your first draft.

   c. Share your definition with other groups.

   d. If your definition or rule has changed, note the change and explain why you like this version better than your original version.

Trapezoids

3. **a.** Make as many different trapezoids as you can, using pieces from one tangram set. Make a copy of each trapezoid by tracing it.

   b. Compare your results with those of another group (or of the whole class). Add any new drawings of trapezoids to your group's collection.

4. **a.** Separate the trapezoids into subsets so that all the trapezoids in a subset are similar to each other.

   b. Use your definition or rule from Step 2 to determine which trapezoids are similar. If it still works, move on. If it doesn't, explain why, and then modify your definition or rule.

   c. Share your definition with other groups.

   d. If your definition or rule has changed, note the change and explain why you like this version better than your original version.

Extensions

5. Can you extend your definition of similarity to any geometric figure? Make a tangram figure that is not a triangle or a quadrilateral. Now make a similar figure on another sheet of paper. Explain how you did it and justify your solution.

6. Irma says that all of the triangles that you can make with tangrams will be similar to one another. What do you think of Irma's conjecture? Explain your reasoning.

*Mathematics
for Elementary
School Teachers*
p. 545

EXPLORATION 9.3   Polyomino Explorations

PART 1: Slides, flips, and turns with pentominoes

1.   There are four pairs of pentomino figures below. In each case, the top figure was transformed into the bottom figure by a flip or a turn. Describe what was done to the top figure to create the bottom figure. Use your pentomino pieces to check your answer.

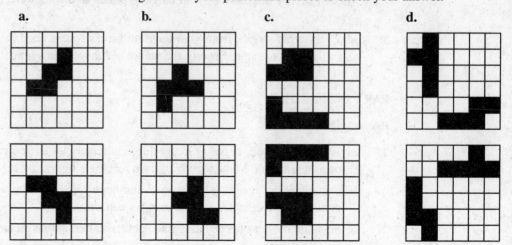

a.   b.   c.   d.

2.   Describe what was done to the top figure to create the bottom figure.

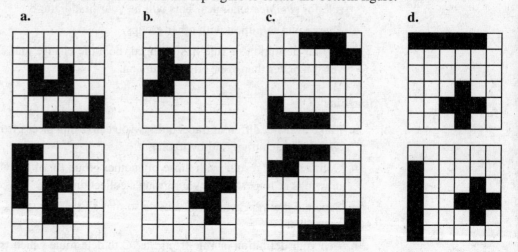

a.   b.   c.   d.

3.   Use the grids on page 253. Make your own pentomino figures on the top grid. Then transform the figure by making any combination of flips or turns. Record the result on the bottom grid. Describe what you did. Give your problems to a friend and see whether your friend can guess your transformations.

Grids for EXPLORATION 9.3,
PART 1: Slides, flips, and turns with pentominoes

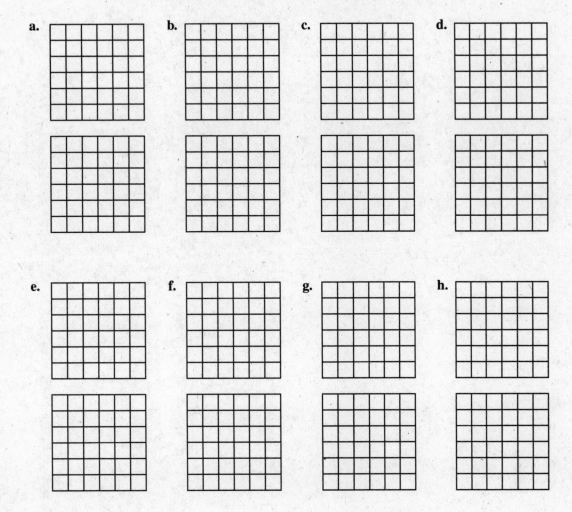

PART 2: Tessellations

1. Which of the five tetromino pieces will tile the plane? Show your solutions. *Note*: Some of the pieces will tile the plane in more than one way; that is, they will make different patterns.

2. Make a tessellation pattern with two or more of the tetromino pieces.

PART 3: Symmetry

Describe the symmetries of each of the 12 pentomino pieces.

PART 4: Similarity

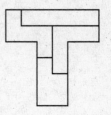

Select the pentominoes that are shaped like the following capital letters: T, U, Z, W, and P.

For each one of these five shapes, a similar figure can be made by using four other pentominoes (of any shape). The solution for T is shown.

1. Take a few minutes to think about making a U, Z, W, or P. What ideas do you have beyond a random guess–check–revise?

2. Share your thoughts with your partners. Together, make a figure similar to one of these pentominoes.

3. At the most basic van Hiele level, the larger figures are similar in appearance to the smaller figures. However, if we analyze the smaller and larger figures, we can find mathematical relationships between the two figures. Take some time in your group to grapple with this concept of similarity, and then draft a definition of similarity that you can then test on these figures.

4. Share your definition with other groups.

5. If your definition has changed, note the changes and explain why you like this version better than your original version.

Mathematics
for Elementary
School Teachers
p. 549

SECTION 9.1   Exploring Translations, Reflections, and Rotations

EXPLORATION 9.4   Reflections (Flips)

PART 1: Developing reflection sense

1.   Take out page 259. For each of the figures at the top of the page, do the following:

    (i)   Sketch where you think the reflection of each shape will be.

    (ii)   Then reflect the shape according to your instructor's directions.

    (iii)   Compare the actual reflection and your prediction. If your prediction was not close, take a moment to analyze your error before going to the next problem.

2.   After doing Step 1, describe what generalizations you can make about reflecting any figure across any line.

3.   For each pair of figures at the bottom of page 259, do the following:

    (i)   Sketch where you think the line of reflection is for each pair of shapes.

    (ii)   Determine the actual line by folding the paper.

    (iii)   Compare the actual line and your prediction. If your predicted line was not close, take a moment to analyze your error before going to the next problem.

4.   After doing Step 3, describe what generalizations you can make about how to determine the line of reflection for a figure and its reflection.

Part 2: Two reflections

Now that you have examined the operation of reflecting a figure across a single line, let us examine the operation of reflecting a figure across two lines in turn.

1.   We will first explore the effect of reflecting a figure across two lines that are parallel.

    a. Look at page 261. Sketch your prediction of where the reflection image of trapezoid *STAR* will be if we reflect it across line *l* and then reflect that image across line *m*. We will use this notation to describe these directions: $STAR(R_l R_m)$.

    b. Do the actual reflections and revise your thinking if your prediction was off.

    c. Now predict the reflection image of $STAR(R_m R_l)$. Briefly explain your prediction.

    d. Do the actual reflections and revise your thinking if your prediction was off.

2.   When mathematicians wish to investigate a phenomenon, they often work systematically. Let us explore another case: What if the figure is between the two parallel lines? In this case will $F(R_m R_l)$ and $F(R_l R_m)$ be equivalent?

    a. Predict the reflection image of $PONY(R_l R_m)$. Briefly explain your prediction.

    b. Do the actual reflections and revise your thinking if your prediction was off.

    c. Predict the reflection image of $PONY(R_m R_l)$. Briefly explain your prediction.

    d. Do the actual reflections and revise your thinking if your prediction was off.

3.   **a.** On a sheet of blank paper make a simple geometric figure *F* and two parallel lines *l* and *m*. Find the reflection images of $F(R_l R_m)$ and $F(R_m R_l)$.

    b. Repeat this process with other figures until you feel confident about answering (c).

    c. What generalizations can you make about the effect of reflecting a figure across two parallel lines?

4. Do you think there is any scenario in which $F(R_m R_l)$ and $F(R_l R_m)$ will be equivalent when reflecting across two parallel lines? Explain your reasoning.

5. Let us now examine reflection across perpendicular lines.

    a. Predict the reflection image of $COLT(R_l R_m)$. Briefly explain your prediction.

    b. Do the actual reflections and revise your thinking if your prediction was off.

    c. Predict the reflection image of $COLT(R_m R_l)$. Briefly explain your prediction.

    d. Do the actual reflections and revise your thinking if your prediction was off.

6. In this case, we found that reflection of this figure across these two perpendicular lines was commutative. Do you think this will always be the case? For example, what if the base of the figure was not parallel to one of the lines of reflection?

    a. Describe your initial hypothesis and briefly explain your reasoning.

    b. Do a number of reflections on your own.

    c. What generalizations can you make about the effect of reflecting a figure across two perpendicular lines?

7. Is there any other transformation or combination of transformations that could achieve the same result as reflecting a figure across two parallel lines? State your thinking and support your statement.

8. Is there any other transformation or combination of transformations that could achieve the same result as reflecting a figure across two perpendicular lines? State your thinking and support your statement.

Figures for EXPLORATION 9.4, PART 1: Developing
reflection sense

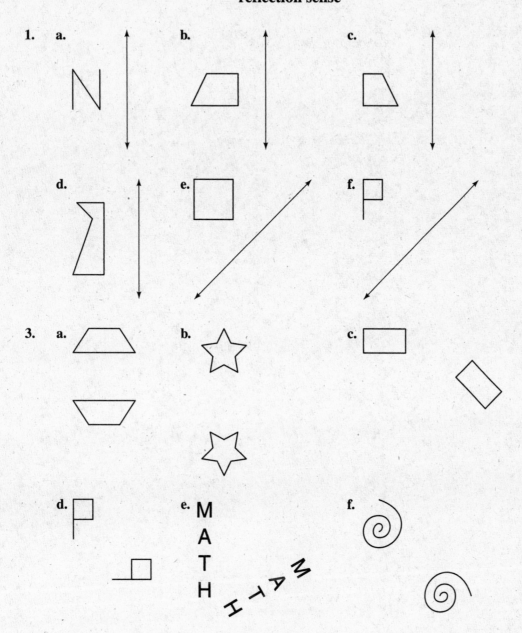

Figures for EXPLORATION 9.4, PART 2: Two reflections

1.

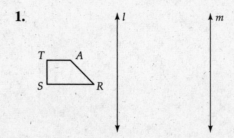

3.

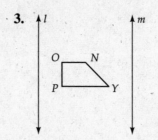

5.

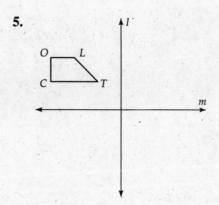

*Mathematics
for Elementary
School Teachers*
p. 549

EXPLORATION 9.5 Paper Folding

Begin with square sheets of paper.

PART 1: Making geometric figures with paper folding

1. ***Square*** Cut two strips of paper of equal width. Fold each strip onto itself. Insert one strip inside the other so that they interlock, as shown at the right. Cut off the excess paper, and you have a square (actually four squares if you unfold what remains and cut at the folds). Why are these figures squares?

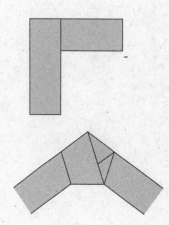

2. ***Pentagon*** Cut a long, thin strip of paper. (This exercise will work with a thin strip cut from $8\frac{1}{2}$ by 11 inch notebook paper). Tie a regular overhand knot with the paper and carefully tighten the knot until your paper looks like the figure at the right. You may have to play with it for a bit. When you cut off the two excess pieces, you have a regular pentagon. What can you see from the folds that can help you to understand why the resulting figure is a regular pentagon?

3. ***Hexagon*** Cut two thin strips of paper. Again, this will work with strips cut from $8\frac{1}{2}$ by 11 inch notebook paper. Tie a square knot with the two pieces of paper. The figure at the left below shows the parts of each strip that are above and below the other strip. Be patient—it takes a couple readings to figure this out. Carefully tighten the knot; as you begin to tighten the knot, your strips should look like the figure at the right. (*Hint*: Make sure that the left and right end pieces are on top of each other as shown in the figure at the left.) What can you see from the folds that can help you to understand why the resulting figure is a regular hexagon?

4. *A family of fives*

    a. Take a sheet of notebook paper and fold it in half as shown in Step 1.

    b. Find the midpoint of the bottom edge by folding. Bring the top left corner of the paper to the midpoint and fold, as shown in Step 2.

    c. Fold the bottom left corner across the fold line as shown in Step 3.

    d. Finally, fold the top edge across the fold line as shown in Step 4. Your result should look like the righthand figure in Step 4.

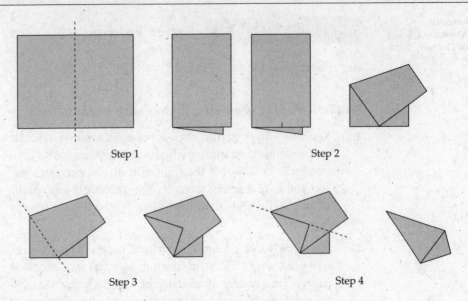

Step 1   Step 2

Step 3   Step 4

e. Cut off the tip at an angle and then open up your paper.

f. Repeat steps (a) through (e), making cuts of various sizes and angles at the tip.

g. Present your results. Explain why each cut produces the kind of figure that it produces.

PART 2: Predicting what happens when you unfold the paper[1]

1. In the example shown below, a piece of paper is folded twice, and then an arrow is drawn in the upper right corner. This is a special type of paper: When the arrow is drawn, its image appears on each of the three layers below. When the paper is unfolded, the four arrows appear as shown.

Example:

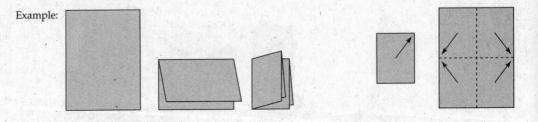

In each part, select the diagram (1, 2, or 3) that shows what the given paper will look like when it is unfolded. Briefly summarize your thinking process.

a.   **b.**

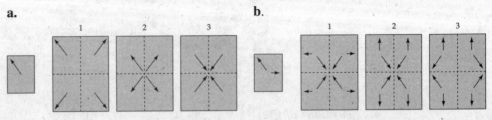

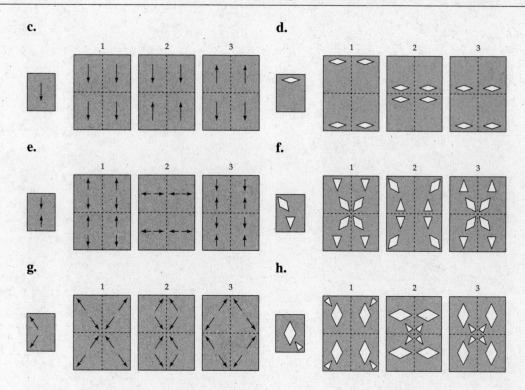

2. Use the diagrams on page 267. In each part, a square piece of paper is folded and cut as shown in the left figure. In parts (a) through (d), the square paper is folded in half once. In parts (e) through (h), the square piece of paper is folded in half twice. The dotted lines represent the fold lines. In each case, use the right figure to draw what the design will look like when the paper is unfolded. Briefly summarize your thinking process.

Figures for EXPLORATION 9.5,
PART 2: Predicting what happens when you unfold the paper

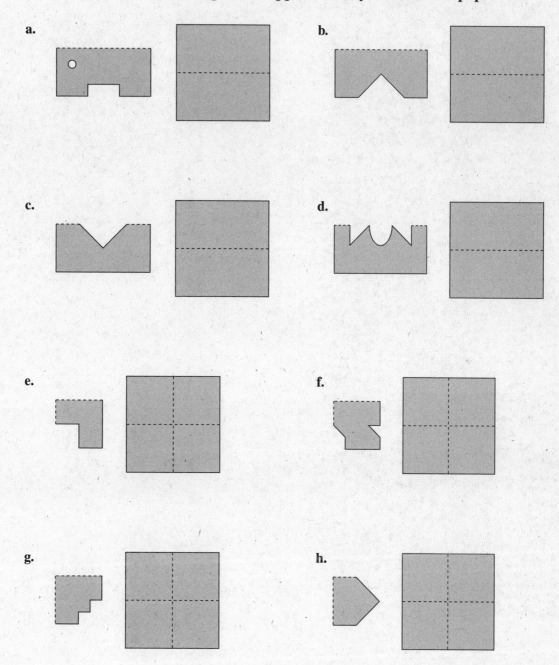

PART 3: Making copies of designs

For each of the questions in this part, whether you produce the design on your first attempt or after several attempts, turn in your solution and a summary of your thinking process. In cases of more than one attempt, include your "failures" and a description of what you learned from each "failure." Many students say they learned a lot from their wrong answers.

1.   Fold a square piece of paper in half. Determine the cut(s) needed to produce each design when the paper is unfolded.

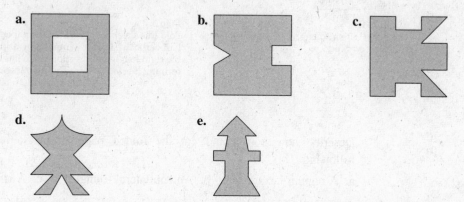

2.   Fold an $8\frac{1}{2}$ by 11 inch sheet of paper in half and then in half again, as shown.

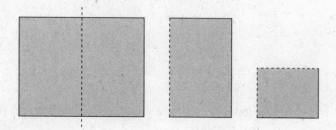

Determine the cut(s) needed to produce each design when the paper is unfolded.

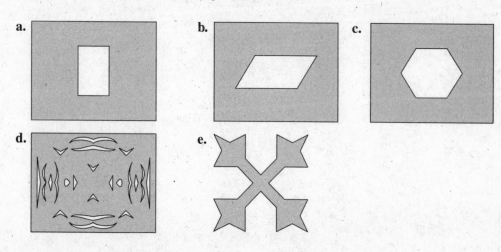

3. Perform the following steps on a square piece of paper. If all goes as planned, you should end up with the paper containing, when unfolded, six 60-degree angles.

  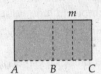

Step 1.
Begin with a square sheet of paper.

Step 2.
Fold the sheet in half.

Step 3.
Fold this sheet in half vertically. Fold the right half in half, forming line *m*.

Step 4.
Fold the paper at point *B* by bringing point *A* to line *m*. This guarantees that angle *ABC* = 60°. Why?

Step 5.
Fold the paper along *AB*, bringing point *C* over so that it lies on top of the left edge of the paper.

Describe how you would cut the folded paper from Step 5 to make each of the following.

a. A regular hexagon   b. An equilateral triangle   c. A six-pointed star

d.    e.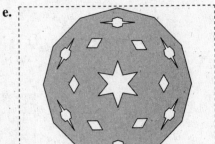

f. Find a picture of a snowflake on the Internet or in a book and make a paper replica of that snowflake.

Mathematics for Elementary School Teachers
p. 550

EXPLORATION 9.6 Developing Rotation Sense

1. For each of the figures on page 273, do the following:

    (i) First, predict the image when the figure is rotated 180 degrees clockwise about the dot by lightly sketching your prediction in pencil on the sheet.

    (ii) Then do the rotation according to your instructor's directions or by the following method. Take a blank sheet of paper and trace each figure and dot (or simply make a copy of the page). Place one sheet above the other. With your pencil or pen firmly on the dot, turn the bottom sheet 180 degrees. Now trace the image.

    (iii) Compare the actual rotation and your prediction. If your prediction was very close, great! If it was not, take a moment to analyze your error. Why was your prediction wrong, how wrong was it, and what can you learn about the relationship between a figure and its rotated image? Write down your analysis!

2. After doing all 9 rotations, describe your method for determining a rotated image. That is, what generalizations can you make about rotating any figure 180 degrees clockwise? *Note*: You are welcome to make up new figures and do more exploring before answering this question.

3. Study the first six rotations (where the dot was not on the figure). What commonalities do you observe between the figure and its rotated image that are true in all six cases?

4. Repeat Steps 1–3 with a 90-degree clockwise rotation.

Figures for EXPLORATION 9.6: Developing Rotation Sense

a.

b.

c. A

d.

e.

f. A

g.

h.

i. A

SECTION 9.2   Symmetry and Tessellations

Symmetry is one of those mathematical concepts that we find virtually everywhere! We find it in the natural world—at the microscopic level and at the galactic level. We find it in plants and animals. We find it in everyday life and in the arts. In some cases it has aesthetic value, in other cases very practical value. We find it in every human culture ever known!

Mathematics for Elementary School Teachers pp. 566, 567, 568

EXPLORATION 9.7   Symmetries of Common Polygons

1.   Name each shape and determine its rotation and reflection symmetries. Describe any questions or problems you have.

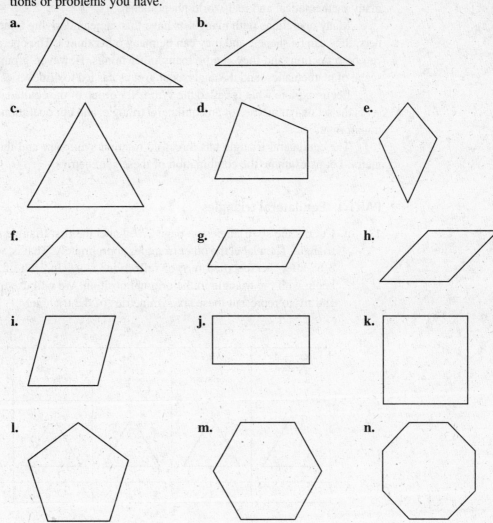

a.

b.

c.

d.

e.

f.

g.

h.

i.

j.

k.

l.

m.

n.

2.   Discuss how to communicate your findings from Step 1.

3.   Make up and test your own figures for rotation and reflection symmetry.

EXPLORATION 9.8 Symmetry Groups

This exploration seeks to deepen your understanding of the connection between operations on shapes and operations on numbers. Let us first review what we know about one operation on one set of numbers: addition on the set of integers.

- The operation of addition is associative; that is, $(a + b) + c = a + (b + c)$.
- There is an identity (0) for that operation.
- There is an inverse for every number.

When a set (in this case, a set of numbers) and an operation satisfy these three criteria, we call the relationship a group. Groups are critically important in understanding the underlying structure of many mathematical and real-world phenomena.

Many operations with many sets have this same underlying structure. The sets can be numbers, they can be shapes, and they can be many other things. The operations can be addition, they can be reflection, and they can be many other things. However, groups appear all over the landscape of mathematics, and the analysis of groups has led to discoveries in science and other fields.

Let us explore what is called the symmetry group of the equilateral triangle. In this case, our set is the set of symmetries of an equilateral triangle, and our operation is the composition of those transformations.

The equilateral triangle has threefold rotation symmetry and three lines of reflection symmetry. Let us examine the combination of these symmetries.

PART 1: Equilateral triangles

1. **a.** Cut out the six triangles on page 279. Label the first triangle as shown in the bottom left triangle. Then label the other triangles appropriately. That is, where will vertices 1, 2, and 3 be after the specified turn or reflection? *Note*: In each of the five cases, you always begin with the triangle in the original position. We will use the symbols I, r, r_2, m_1, m_2, and m_3 to represent these six symmetries of the triangle.

Original position	After $\frac{1}{3}$ turn clockwise	After $\frac{2}{3}$ turn clockwise	After vertical reflection	After this diagonal reflection	After this diagonal reflection
I	r	r_2	m_1	m_2	m_3

b. Fill in the "multiplication" table that appears below. In each case, you begin in the original position, so the top left cell represents the identity operation—that is, do nothing. This is like 1×1 in multiplication.

Look at r times r. That is, if you perform a $\frac{1}{3}$ turn and then another $\frac{1}{3}$ turn, the triangle will be in the same position as if you had done a $\frac{2}{3}$ turn. Thus, r times r = r_2.

One more example: Do a $\frac{1}{3}$ turn and then do reflection m_1—that is, vertical reflection. What do you get? If you did it correctly, you found that this composition is equal to doing the m_3 reflection to the original figure.

△	I	r	r_2	m_1	m_2	m_3
I	I					
r		r_2		m_3		
r_2						
m_1						
m_2						
m_3						

c. Compare results with a partner to make sure you are accurate.

d. What do you see? (This is an intentionally wide-open question.)

e. Write down all the patterns you see in the table.

f. Select one nontrivial pattern to describe carefully. Exchange descriptions with a partner. If your partner was able to understand the pattern, great. If not, circle the words or phrases that caused the problem. Then rewrite and exchange with another partner.

g. Do this set and this operation satisfy the definition of a group?

  (1) Is there an associative property $(a \times b) \times c = a \times (b \times c)$? What is your support for your answer?

  (2) Does each operation have an inverse? If so, write the inverse of each operation.

$$I^{-1} = \underline{\quad} \quad r^{-1} = \underline{\quad} \quad r_2^{-1} = \underline{\quad} \quad m_1^{-1} = \underline{\quad} \quad m_2^{-1} = \underline{\quad} \quad m_3^{-1} = \underline{\quad}$$

2. Careful examination enables us to realize that we don't *need* three different mirrors. That is, the m_2 reflection is equivalent to the vertical reflection and a $\frac{1}{3}$ turn. Thus the m_2 reflection is equivalent to the m_1 and r composition. Similarly, the m_3 reflection is equivalent to the vertical reflection and a $\frac{2}{3}$ turn. This discovery simplifies our notation, because we need only specify one line of reflection: vertical. And so the new multiplication table for the symmetries of an equilateral triangle looks like the table below. That is, analysis of our operations enables us to change our notation to reflect the relationships/connections that we found. Fill in the table below.

△	I	r	r_2	m	mr	mr_2
I						
r						
r_2						
m						
mr						
mr_2						

PART 2: Squares

For this part, you will need to cut out the eight squares on page 279.

1.   **a.** Label the first square as shown to the right. Then label the other squares appropriately. That is, where will vertices 1, 2, 3, and 4 be after the specified turn or reflection?

1	2
4	3

   b. Fill in the table below to determine the multiplication table for the symmetries of a square.

□	I	r	r_2	r_3	m	mr	mr_2	mr_3
I								
r								
r_2								
r_3								
m								
mr								
mr_2								
mr_3								

   c. Compare results with a partner to make sure you are accurate.

   d. What do you see?

   e. What patterns do you see?

   f. Select one nontrivial pattern to describe. Exchange descriptions with a partner. If your partner was able to understand the pattern, great. If not, circle the words or phrases that caused the problem. Then rewrite and exchange with another partner.

   g. Do this set and this operation satisfy the definition of a group?

   (1) Is there an associative property $(a \times b) \times c = a \times (b \times c)$? What is your support for your answer?

   We know from before that I is the identity operation for this set.

   (2) Does each operation have an inverse? If so, write the inverse of each operation.

Figures for EXPLORATION 9.8, PART 1 and PART 2

PART 1: Equilateral triangles

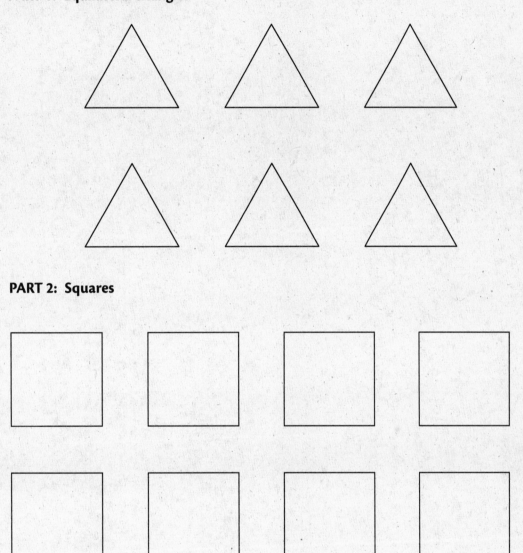

PART 2: Squares

*Mathematics
for Elementary
School Teachers*
p. 579

EXPLORATION 9.9   Tessellations

PART 1: Which figures tessellate?

Look at the following six pairs of figures. Work with several copies of the enlarged versions of the figures in order to determine which of the figures in each pair will tessellate: both, just one, or neither.

If both of the figures tessellate, make and justify a generalization: All figures that "look like" these two and have the following characteristics (describe the characteristics) will tessellate.

If just one of the figures in a pair tessellates, explain why the one does tessellate and why the other doesn't. That is, what characteristics or properties does the one figure have that the other doesn't have?

If neither of the figures tessellates, describe whether you believe that modifications of the figure could be made so that figures that "look like" these two might tessellate.

1.

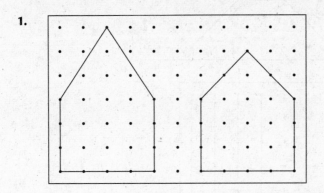

2.

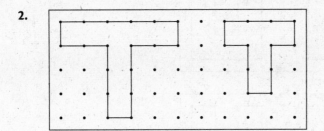

3.

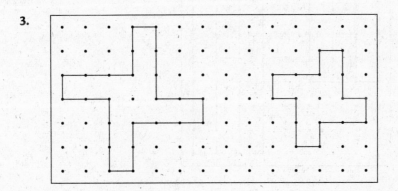

4.

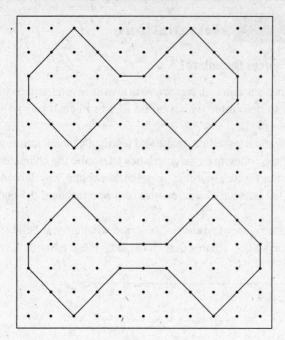

5.

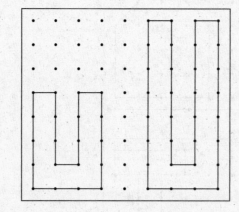

6.

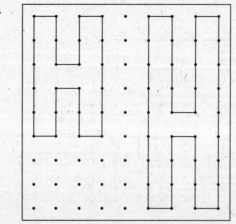

PART 2: What kinds of pentagons will tessellate?

While the regular pentagon does not tessellate, some pentagons *do* tessellate. Under what conditions will a pentagon tessellate? Explore this question by making and testing several different kinds of pentagons.

Tips: Fold a piece of paper in half, then in half again, and then in half again. Now if you draw a pentagon on the page and then cut out the pentagons (with heavy scissors), you will have eight copies of the pentagon. Alternatively, you can make a pentagon using a software program and cut and paste many copies of the pentagon.

Prepare a report that includes

- your conclusion(s) in the following form: A pentagon with these characteristics (describe the characteristics) will tessellate. If you find more than one family that tessellates, write the description of the characteristics of each family.

- your justification of your conclusion(s).

- a brief summary of your solution path: How did you come to your conclusion(s)? This will include your "failures" as well as your "successes."

PART 3: What kinds of arrows will tessellate?

Believe it or not, there are many kinds of arrows that will tessellate. In this exploration, your challenge is to determine the characteristics that are necessary in order for an arrow to tessellate. For example, does it need to be symmetric? Do the sides of the shaft need to be parallel? Can the tip be skinny? Make and test several different kinds of arrows, such as those shown below, either on blank paper or on Geoboard Dot Paper.

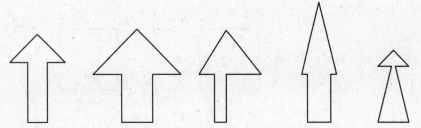

Prepare a report that includes

- your conclusion(s) in the following form: An arrow with these characteristics (describe the characteristics) will tessellate. If you find more than one family that tessellates, write the description of the characteristics of each family.

- your justification of your conclusion(s).

- a brief summary of your solution path: How did you come to your conclusion(s)? This will include your "failures" as well as your "successes."

PART 4: Semiregular tessellations

Now let us expand our discussion to combinations of figures that tessellate. A *semiregular tessellation* occurs when two or more regular polygons tessellate and every vertex point is congruent to every other vertex point.

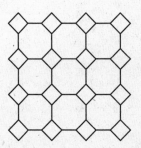

1.   **a.** Using your understanding of congruent, try to define *congruent vertex point.*

  b. Compare definitions with your partner(s). Modify your definition, if needed, as a result of the discussion.

2. Cut out the figures on the Regular Polygons sheet at the end of the book. Use them to explore combinations of simple geometric figures to determine other semiregular tessellations.

    a. Sketch your "successes" and "failures" and record your reflections after each success or failure.

    b. How many semiregular tessellations do you think there are—5, 10, 20, 50, 100, thousands, an infinite number? Explain your reasoning.

3. In your explorations, you may have found some *demiregular tessellations*, which occur when two or more regular polygons tessellate but not every vertex point is congruent to every other vertex point. Show any demiregular tessellations that you have discovered. Explain why they are not semiregular tessellations.

PART 5: Escher-like tessellations

Using the language of translations, reflections, and rotations, we are now in a position to understand how the artist M. C. Escher made his tessellations (see, for example, Figure 9.19 on page 554 of the text). Escher began with a polygon that would tessellate and then transformed that polygon. We can use the analogy of chess to understand how he did this: There are legal moves and illegal moves.

1. **Translations**   One legal move is to modify one side of the polygon and then translate that modification to the opposite side. For example:

    Begin with a square (Step 1).

    Modify one side (Step 2).

    Translate the modified side to the opposite side (Step 3).

    In this case, the translation is a vertical slide.

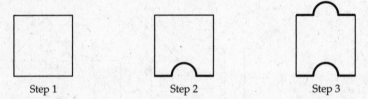

Step 1   Step 2   Step 3

    a. Begin with a square, and experiment with translations. Show your work and reasoning so that your instructor can follow your thinking. Make notes of your observations as they happen: patterns you see, conjectures you decide to pursue, questions you have.

    b. Summarize these observations, patterns, and hypotheses about translations and tessellations.

2. **Rotations**   Another legal move is to rotate part of a shape. The center of rotation—that is, the "hinge" about which the part rotates—can be a vertex or a midpoint of one of the sides. Here is an example:

    Begin with an equilateral triangle (Step 1).

    Modify one side (Step 2).

    Rotate that side 60 degrees counterclockwise about the bottom left vertex of the triangle (Step 3).

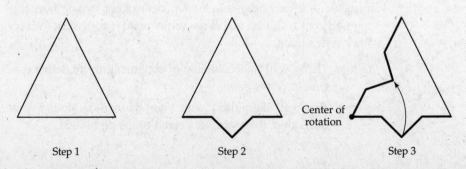

Step 1   Step 2   Step 3

a. Begin with a square or an equilateral triangle, and experiment with rotations. Show your work so that your instructor can follow your thinking. Make notes of your observations as they happen: patterns you see, conjectures you decide to pursue, questions you have.

b. Summarize what you learned about rotations and tessellations.

3.   ***Combinations of translations, rotations, and reflections***   We can combine translations, rotations, and reflections to make tessellations. For example, we can begin with a rectangle (see Step 1 below) and, using a compass and straightedge, construct a semicircle whose diameter is half the length of the side of the rectangle (Step 2).

Further, we can then rotate the semicircle 180 degrees, as shown in Step 3 below, the center of rotation being the midpoint of the top side of the rectangle.

Next, we translate this shape to the bottom side of the rectangle. The result is shown in Step 4. This shape will tessellate, but it's not very interesting.

However, we can also transform the shorter side of the rectangle, as shown in Step 5. This time, we replace the top half of the right side of the rectangle with a quarter-circle whose radius is equal to the length of the line segment we are replacing.

Then we *reflect* (another legal move) that quarter circle across a line that is parallel to the base of the rectangle and goes through the middle of the rectangle. See Step 6.

Finally, we translate this side of the rectangle to the other side, as shown in Step 7. The original rectangle is shown with dotted lines.

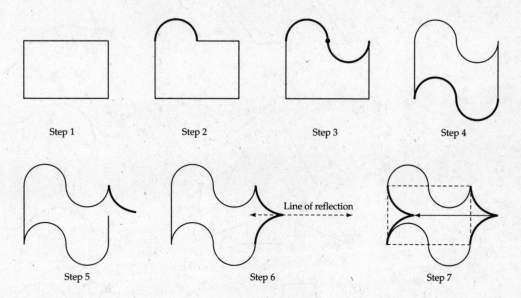

a. Begin with a square or a rectangle, and experiment with combinations of translations, reflections, and rotations. Show your work so that your instructor can follow your thinking.

b. Summarize what you learned about combinations of translations, rotations, and reflections and tessellations.

4.   Each of the figures below tessellates, and each began as a basic geometric figure. Can you determine the starting polygon and how the shape was made?

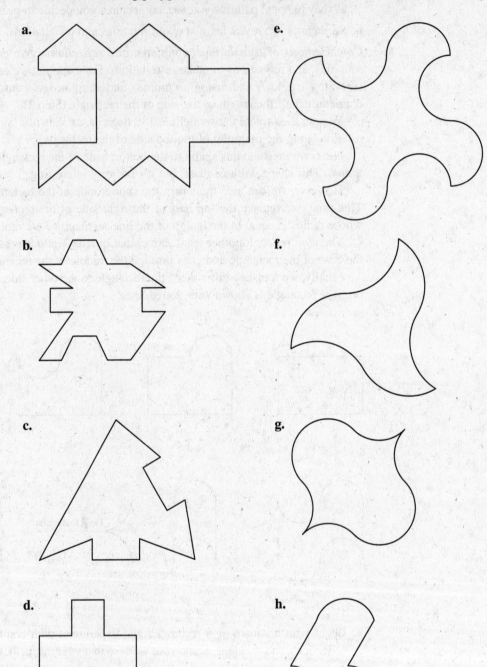

a.

b.

c.

d.

e.

f.

g.

h.

5.   Make your own tessellations!

EXPLORATION 9.10 Quilts

There are so many places to explore symmetry and so little time in this course!

Quilting has a fascinating and long history. Although we cannot definitely mark the date of the first quilt, we know that in the transition from hunter-gatherers to farmers many thousands of years ago, the development of agriculture inspired experiments with new ways to keep warm at night other than using animal hides, leaves, straw, and other such material. The oldest remnants of what we might call quilts—that is, spreads with patterns on them—are well over two thousand years old. Various cultures developed different patterns and techniques. We know that the Crusaders brought the idea of quilting back to Europe from the Middle East. However, an event in the fourteenth century caused quilting to come into its own: Europe suffered through what is referred to as the *Great Freeze*—year after year of very severe winters. This added need for more warmth at night spurred many advances in quilting, including the quilting frame. After the Great Freeze, a much greater percentage of women made quilts, and it was one of the few activities that was engaged in by poor and rich alike!

The origin of what has been called *patchwork quilting* is also interesting. It developed out of pioneer life in the early years of this country's history. As pioneers moved westward, more and more people lived far from stores, and the notion of "having to make do with what you've got" became a part of life. The amount of garbage generated by pioneers was a small fraction of what families throw away today. After making clothes, women would have various pieces of leftover fabric. Rather than throw them away, they would cut the pieces into small triangles, squares, and other geometric figures. The triangle was a common choice because one could get the most from scraps by cutting out triangles. (Why is this?) In many areas, wintertime was the quilting party season because there was little farm work to be done. Women would gather to trade scraps, share ideas, work together, and socialize.

PART 1: Starting simple

The *block* is the basic unit of a quilt. Blocks are usually squares (and we often call them *squares*), but some quilts' blocks are hexagons or other shapes.

1. **a.** Examine the quilt block shown at the left below. This basic unit has been constructed from eight right isosceles triangles using two colors. Describe the construction of this block in terms of translations, rotations, and reflections. You are describing the block's *pattern*, which is sometimes called the *quilt's* pattern. (Quilting terminology is not always precise!)

    b. Examine the shape shown at the right below, which was created by putting together four copies of the quilt block. (A quilt is made by sewing many blocks together.) Describe the new designs and shapes that you see. Explain why these designs came about.

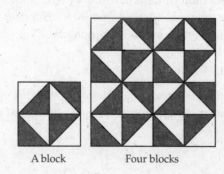

A block   Four blocks

2. Take eight right isosceles triangles and put them together in some way to form a block. Your block does not need to be a square. Sketch your pattern; that is, sketch what your block looks like.

3. Work in groups of four.

    a. Select one person's block and have each member copy that block.

    b. Predict the shapes that will emerge when you put these four congruent blocks together.

    c. Put the four blocks together. Describe the new patterns and shapes that you see. Explain why these patterns came about.

    d. Repeat the process.

    e. Sketch one figure that emerged from combining the four congruent blocks.

    f. Describe where the new patterns came from.

PART 2: What do you see?

1. Your instructor will show a quilt pattern. What do you see? Write your observations.

2. Your instructor will give instructions for drawing a quilt pattern from memory. (*Reminder:* The objective here is not just to get it right but also to identify ideas that will enable you to generate the design.)

    a. Draw the pattern from memory.

    b. Alternatively, describe how you would draw the pattern from memory.

PART 3: Exploring actual quilt patterns

"There are two kinds of people in this world: those who divide everything into two groups and those who don't."[2] We will do this with quilts, for each quilt has many attributes.

1. Take a minute or so for each member of your group silently to examine each quilt pattern on page 291. Then, one by one, describe what you see. That is, member 1 describes what is seen in the first pattern. If member 2 saw something different, that is noted. After each member has noted any additional observations, examine the second pattern. This time member 2 goes first, and so on.

2. Take out page 291 and cut out the quilt blocks. Separate these blocks into two or more groups in such a way that each group has one or more common attributes. Describe the common attribute(s) and name each group. Repeat this as many times as you can in the time provided. (*Note*: Save these 18 quilt blocks for use in Part 7.)

3. What did you learn from classifying the blocks and discussing the classifications?

All quilt patterns have a story. The stories behind some of these quilt blocks are given below. These stories are adapted from *Eight Hands Round: A Patchwork Alphabet*, by Ann Paul, one of many interesting books on quilting.[3]

Storm at Sea

This pattern represents a lighthouse, with the center square representing the light. Before modern technology, lighthouses were built at the most dangerous points along the coast.

Windmill

Windmills used the wind as a source of power so that people could mill their grain. That is, the windmill (and water wheel) would generate enough power to turn a millstone, which would crush the grain and turn it into flour.

Yankee Puzzle

Many years ago, people in New England often played with a puzzle that had seven small pieces—five triangles, one square, and one parallelogram—which they would arrange into different designs.

Quilt Blocks for EXPLORATION 9.10,
PART 3: Exploring actual quilt patterns

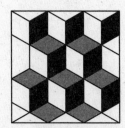

Baby Blocks

Broken Windows

Cross Roads

Does and Darts

Drunkard's Path

Flower Basket

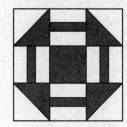

Hole in the Barn Door

Kansas Sunrise

Le Moyne Star

May Basket

Saw Tooth Star

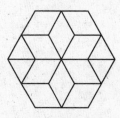

Star

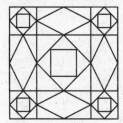

Storm at Sea

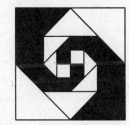

Snail's Trail

Old Tippecanoe

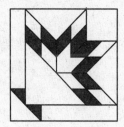

Tulip Basket

Windmill

Yankee Puzzle

PART 4: Closest relative

Look at the quilt blocks below. Find the "closest relative" to Indian Star. Justify your choice. There is no "right" answer. The quality of your answer depends on your justification. This task involves describing the commonalities that you find that make the two patterns so similar. Note that this is a communication task and a reasoning task. How well you do depends partly on applying problem-solving tools and making connections.

Indian Star

25-Patch Star

African Safari

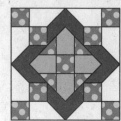

Grand Prix

Hearth and Home

Oregon Trail

Prairies 9 Patch

Right Hand of Friendship

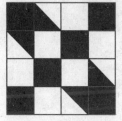

Road to Oklahoma

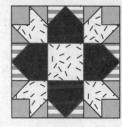

Weathervane

PART 5: Using transformations to construct a quilt pattern

Your instructor will ask you to construct a quilt pattern by beginning with the smallest possible piece(s) and then using slides, flips, and turns to generate the pattern. The example below illustrates one solution path for Dutch Man's Puzzle on page 294.

Begin with the square shown in Step 1. Reflect this square through its right side (Step 2).

Translate this rectangle up one unit (Step 3).

Rotate the figure from Step 3 counterclockwise through a $\frac{1}{4}$ turn (90 degrees); the center of rotation is the top left corner of the figure. The result is shown in Step 4.

Rotate the figure from Step 4 counterclockwise through a $\frac{1}{2}$ turn (180 degrees); the center of rotation is the midpoint of the left side of the figure. The result is shown in Step 5.

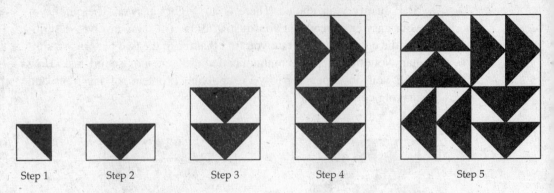

Step 1   Step 2   Step 3   Step 4   Step 5

PART 6: Similarities and differences

For each pair of quilt patterns below, describe how the two patterns are alike and how they are different.

1.

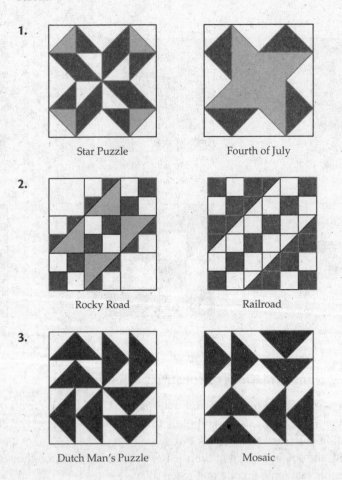

Star Puzzle   Fourth of July

2.

Rocky Road   Railroad

3.

Dutch Man's Puzzle   Mosaic

PART 7: Different symmetries

Take out page 297 and cut apart the 12 quilt patterns. Combine these patterns with the set of quilt patterns in Part 3.

1. Determine which patterns have rotation symmetry. Classify them according to the kind of rotation symmetry that they have.

2. Determine which patterns have reflection symmetry. Classify them according to the kind of reflection symmetry that they have.

Here are the stories behind several of the quilt patterns on page 297.

Anvil

In earlier times, most towns had a blacksmith who made things from iron by placing the metal in a fire until it became soft enough to have its shape changed by pounding it with a hammer against an anvil. The blacksmith made axes, farm tools, horseshoes, and other things that had iron components. The anvil had both a straight part and a curved part. When the blacksmith wanted to make the iron straight, he placed the iron on the flat part of the anvil. When he wanted to make the iron curve, he placed it on the curved part of the anvil.

Eight Hands Round

At various times, people in an area would gather together: for a barn raising, a quilting bee, or some other occasion. In the evening, there would usually be some music and some dancing—square dancing and/or contra dancing. Someone would "call" the dances—that is, give instructions throughout the dance. One of the calls was "eight hands round," which meant for the four couples in a square to join hands in a circle.

Log Cabin

When the American settlers moved to the frontier, they had to make their own homes from the wood in the forest. You may have heard the stories about Abraham Lincoln's being born in a log cabin. This pattern was inspired by someone's creative representation of the pattern of the logs in a log cabin.

Underground Railroad

The Underground Railroad was a network of people who helped slaves escape from slavery. A slave trying to get to a state where slavery was illegal often had to travel hundreds of miles. Those people who helped the slaves fed and hid them, usually at night, and then gave them directions to the next safe place.

3. Using different colors of construction paper, cut out a number of squares and triangles.
   a. Make a pattern (a block) that has rotation symmetry but not reflection symmetry. Briefly convince the reader that the figure does have rotation symmetry but not reflection symmetry.
   b. Make a pattern that has reflection symmetry but not rotation symmetry. Briefly convince the reader that the figure does have reflection symmetry but not rotation symmetry.
   c. Make a pattern that has both reflection symmetry and rotation symmetry. Briefly convince the reader that the figure does have rotation symmetry and reflection symmetry.

Quilt Patterns for EXPLORATION 9.10, PART 7: Different symmetries

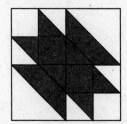

Anvil

Crazy Ann

Dutch Man's Puzzle

Eight Hands Round

Flying Geese

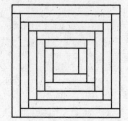

Log Cabin

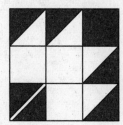

Maple

Peony

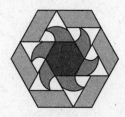

Starry Nights

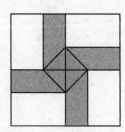

Susannah

Rabbit Paw

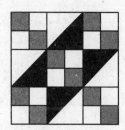

Underground Railroad

SECTION 9.3   Exploring Similarity

Similarity with Pattern Blocks

In Chapter 5, we used Pattern Blocks to explore fraction concepts. These blocks also let us see geometry "in action."

PART 1: Triangles

1. Take the small green triangle. Using the Pattern Blocks, make a bigger triangle that is similar in shape to this triangle. Explain why you believe this triangle is similar to the original triangle.

2. Make another triangle that is similar to the original triangle. Explain why you believe this triangle is similar to the original triangle.

3. Record other observations: patterns, conjectures, and questions. For example, are there patterns in the lengths of the sides? In how large the figures are? In how you actually make larger figures?

PART 2: Squares

1. Take the small orange square. Using the Pattern Blocks, make a bigger square that is similar in shape to this square. Explain why you believe this square is similar to the original square.

2. Make another square that is similar to the original square. Explain why you believe this square is similar to the original square.

3. Record other observations: patterns, conjectures, and questions. For example, are there patterns in the lengths of the sides? In how large the figures are? In how you actually make larger figures?

PART 3: Parallelograms

As it turns out, all equilateral triangles are similar to one another, and all squares are similar to one another. This is not true for all parallelograms!

1. Take the small blue parallelogram. Using the Pattern Blocks, make a bigger parallelogram that is similar in shape to this parallelogram. Explain why you believe this parallelogram is similar to the original parallelogram.

2. Make another parallelogram that is similar to the original parallelogram. Explain why you believe this parallelogram is similar to the original parallelogram.

3. a. Write your first draft of a definition of *similar* with respect to parallelograms. Your definition should be more precise than "same shape, possibly different size."

    b. Compare your definition with that of your partner(s). After the discussion, modify your definition, if needed.

4. Record other observations: patterns, conjectures, and questions. For example, are there patterns in the lengths of the sides? In how large the figures are? In how you actually make larger figures, and so on?

PART 4: Trapezoids

1. Take the small red trapezoid. Using the Pattern Blocks, make a bigger trapezoid that is similar in shape to this trapezoid. Explain why you believe this trapezoid is similar to the original trapezoid.

2. Make another trapezoid that is similar to the original trapezoid. Explain why you believe this trapezoid is similar to the original trapezoid.

3. Record other observations: patterns, conjectures, and questions. For example, are there patterns in the lengths of the sides? In how large the figures are? In how you actually make larger figures, and so on?

4. Which of the following trapezoids is similar to the small red trapezoid? Justify your reasoning.

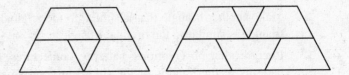

5. a. Write your first draft of a definition of *similar* that would be true for any geometric figure. Be precise, as noted in Part 3.

    b. Compare your definition with that of your partner(s). After the discussion, modify your definition, if needed.

Extension

There is a pattern concerning the area of consecutive similar pattern blocks that is true for all of the blocks: The ratio of the area of the next bigger similar figure to the original is 4:1. That is, if we count the area of the green triangle as 1 unit, the next bigger equilateral triangle has an area of 4 units. The blue parallelogram has an area of 2 units, and the next bigger blue parallelogram has an area of 8 units. Thus the ratio of their areas is 4:1. This is also true for the orange square, the red trapezoid, and the yellow hexagon.

   Do you think this is true only in these cases, or will it be true for composite figures too? For example, will the area of the next bigger figure similar to this one be 4 times as great as the area of this one?

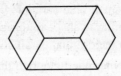

*Mathematics
for Elementary
School Teachers*
p. 596

EXPLORATION 9.12   Similar Figures

PART 1: Similar triangles

In high school geometry, we discovered that we did not have to show that all six pairs of corresponding angles and sides were congruent in order to know that two triangles were congruent. We learned about SSS, SAS, ASA, AAS, and HL. What might we need to show in order to know that two triangles are similar? At this point, in order to know that the two triangles are similar, we need to measure the lengths of all six sides and all six angles, and then we have to determine three ratios. If all the ratios are equal and all three pairs of angles are equal, then we know the two triangles are similar.

$$\frac{AB}{PQ} = \frac{AC}{PR} = \frac{BC}{QR} \quad \text{and} \quad m \angle A = m \angle P, \; m \angle B = m \angle Q, \; m \angle C = m \angle R$$

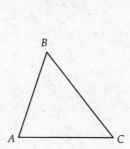
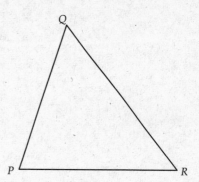

The question is: Do we have to know all the ratios to determine whether two triangles are similar? What combinations of sides and angles will be sufficient?

1.   Write your initial hypothesis and reasoning about *one* combination that is sufficient to prove that two triangles are similar.

2.   Discuss your ideas with your partner(s). Take one hypothesis; try to convince yourself that it is true, and try to make triangles that are not similar but that satisfy your hypothesis.

3.   Present your hypothesis to another group. Note any comments or suggestions made by the other group.

4.   Work through Part 1 repeatedly (as time permits) in order to convince yourself of as many similarity combinations as you can.

PART 2: Similar quadrilaterals

The figure below shows a pair of similar rectangles and a pair of similar trapezoids. At this stage of our explorations, to confirm that they are indeed similar, we would have to measure all eight sides and all eight angles and verify that the lengths of corresponding pairs of sides had the same ratio and that corresponding angles were congruent. The question is: Do we have to know all eight pieces of information? What combinations of sides and angles will be sufficient?

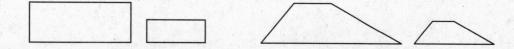

1. *Rectangles*

    a. Write your initial hypotheses and reasoning for rectangles.

    b. Discuss your ideas with your partner(s). Take one of your hypotheses; try to convince yourself that it is true, and try to make rectangles that are not similar but that satisfy your hypothesis.

    c. Present your hypotheses to another group. Note any comments or suggestions made by the other group.

2. *Trapezoids*   Repeat Step 1 for trapezoids.

Extension

Come up with a conjecture that will work for all quadrilaterals. That is, if these conditions are met, then the two quadrilaterals will be similar. Justify your conjecture.

CHAPTER 10

Geometry as Measurement

M any, if not most, uses of basic mathematics involve measurement. Formulas are important because they enable us to answer the "how much?" questions more efficiently. However, for many problems, there is no formula (for example, what is the area of your state?). In other cases, you need to apply measurement ideas to solve nonroutine problems (for example, to find the height of a tree). In this chapter, you will explore important measurement ideas and also come to understand why various formulas work.

Exploring Systems of Measurement

In this section, we will examine some of the issues surrounding linear measurement—that is, situations in which we want to ask questions or gain information about distances, heights, and thicknesses. We will also work with measuring weight and begin thinking about area and volume.

Mathematics for Elementary School Teachers p. 605

EXPLORATION 10.1 **How Far Is It?**

Before modern measuring methods evolved, people determined distances by using their pace as the unit. Armies needed to know how far they had gone; settlers needed to know how close the nearest house was, and so on. We will replicate that process in this exploration.

1. Determine two points on campus for which you will approximate the distance.

2. Discuss how you will determine the length of one of your paces.

   a. Discuss what one pace means and how you might measure one pace.

   b. Determine how you will measure one pace.

   c. Practice walking so that your paces will be as even as possible.

3. Collect three measurements of your pace. If all three are close, fine. If they are not, decide whether you will take the average of the three or will practice some more and then try again.

4. Walk the distance determined in Step 1 at least one time, and determine that distance (in either feet or meters) on the basis of your pace.

5. Collect the data for the whole class.

6. Analyze those data.

   a. Find the mean, median, and mode. If you had to pick one number, which would you pick? Why?

   b. Make one graph for the data. Describe how the graph adds to your understanding of the data and tell how it helps us to answer the question "how far is it?"

   c. What other analyses of the data would you present if you were reporting this distance?

7. Suppose you were reporting this distance to someone for whom this information was very important (such as an army general or the leader of the settlement). Judging on the basis of the data, what do you think is the distance? Discuss the level of your confidence in this number.

8. If you have a trundle wheel or other measuring device, determine the actual distance. How close was your approximation?

 You can see third-graders' thinking about determining their pace in *Measuring Space in One, Two, and Three Dimensions: Casebook,* by D. Schifter, V. Bastable, and S. J. Russell with K. R. Woleck (Parsippany, NJ: Dale Seymour Publications, 2002), pp. 71–77.

Mathematics for Elementary School Teachers
p. 605

EXPLORATION 10.2   How Tall?

In many cases, we obtain a linear measurement without measuring the object directly. For example, we might read that the distance from New York to Los Angeles is 2825 miles and that the Willis Tower in Chicago is 1454 feet tall. These numbers were not determined using tape measures! What do you do when you cannot measure something directly?

1. Your instructor will select a tree or building for which you will determine the height.

    a. Brainstorm reasons why someone might actually want to know the height of this object.

    b. Brainstorm ideas for determining the height.

2. Select and pursue one method with your group. Describe and justify each step in your plan.

3. **a.** Justify the precision in your answer—that is, the choice of unit and the degree of accuracy: for example, to the nearest 10 feet (meters), 1 foot (meter), one decimal place, and so on.

    b. Describe and explain your degree of confidence in your result.

    c. Describe any difficulties you had and how you overcame them.

4. Present your group's findings to the class.

5. Which group's method do you think will produce the answer closest to the actual height? Justify your response.

You can see some children's strategies for solving this problem in "Third-Grade Students Engage in a Playground Measuring Activity" in the November 1997 issue of *Teaching Children Mathematics*.

*Mathematics
for Elementary
School Teachers*
p. 605

EXPLORATION 10.3 How Thick?

Most of our linear measurements use units such as inches, feet, or miles, or the corresponding metric units. However, people in many occupations need measurements of amounts that are very small. For example, how thick is your skin? What is the wavelength of red light? How thick is an amoeba?

Let us explore how we might answer one such question: How thick is one sheet of paper?

1. With your partner(s), brainstorm ideas for answering this question.

2. Select and pursue one method with your group. Describe and justify each step in your plan.

3. a. Justify the precision in your answer—the choice of unit and the degree of accuracy, for example.

   b. Describe and explain your degree of confidence in your result.

   c. Describe any difficulties you had and how you overcame them.

4. Present your group's findings to the class.

5. Which group's method do you think will produce the answer closest to the actual thickness? Justify your response.

EXPLORATION 10.4 How Much Is a Million?

Researchers have told us that many people's behavior with numbers is very different when the size of the numbers becomes greater than they can actually imagine. For example, we can concretely imagine how much space would be taken up by 100 people listening to a musician, but it is hard to imagine how much space would be taken up by a concert audience of 1 million people.

Many elementary school teachers have told me that when they ask, for example, "How large does a container need to be to hold one million paperclips?", children invariably underestimate. Many give responses such as "a jewelry box" or "a cereal box."

For each of the questions below, do the following:

a. Write down your estimate and your reasoning behind that estimate.

b. Devise a plan for answering the question.

c. Present your results.

PART 1: 1 million dollars

Betty has just won 1 million dollars in 1-dollar bills!

1. How much do you think 1 million 1-dollar bills will weigh?

2. What if we made a road with these dollar bills, in which the width of the road consisted of 10 1-dollar bills? How long would the road be?

3. What if we were to stack these dollar bills in a room? What would be the dimensions of a container that would hold 1 million 1-dollar bills?

PART 2: 1 million pennies

1. How much do you think 1 million pennies would weigh?

2. In *The Wizard of Oz*, Dorothy walked down the yellow brick road. What if we made a road that was 2 feet wide with 1 million pennies? How long would the road be?

3. What if we were to stack these 1 million pennies in a room, after first putting groups of 50 pennies into rolls? What would be the dimensions of a room that would hold 1 million pennies?

SECTION 10.2   Exploring Perimeter and Area

We encounter the need to know perimeters and areas in our everyday and work lives. In determining perimeters and areas, we need many ideas and formulas that have been developed over the centuries.

*Mathematics
for Elementary
School Teachers*
p. 620

EXPLORATION 10.5   What Does π Mean?

1. What does π mean? Suppose someone asked you this question. The person knows that the value of π is approximately 3.14 but does not have a sense of what π means. How would you answer that question? Write your first-draft thoughts before reading on.

2. After observing your instructor's demonstration or doing explorations provided by your instructor, describe what π means (second draft).

3. In groups of three, go through the following process:

    a. Each person reads her or his response. The others give feedback with respect to the accuracy and clarity of the response, in that order.

    • Accuracy: If a group member feels that the statement is not entirely accurate, discuss this issue until it is resolved.

    • Clarity: If a group member feels that certain words or phrases are ambiguous or unclear or vague, discuss those issues until they are resolved.

    b. Write down your third draft of "what π means."

4. Describe any questions you have about π at this point.

*Mathematics
for Elementary
School Teachers*
p. 621

EXPLORATION 10.6 Exploring the Meaning of Area

What Does Area Mean?

1. Let us begin by exploring what area means—not how we determine it, but what it means. Imagine that someone from another planet came to visit and that when you made a comment like "The area of this figure is greater than the area of that figure," the extraterrestrial asked you what *area* means. How would you respond? Write your response and then compare responses with your partner(s).

What Does the Number Mean?

2. Consider a rectangle whose dimensions are 12 inches by 24 inches. We might say that the area is 288 square inches, or, if we used feet as our unit of measurement, we might say that the area is 2 square feet. If we were in a country that used the metric system of measurement, we would determine the length and width to be approximately 60 centimeters and 30 centimeters, and we would say that the area is 1800 square centimeters. From this perspective, there is not a functional relationship between an object and the number used to denote its area, unless you specify the unit of measurement.

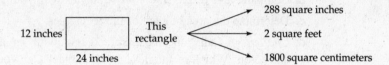

12 inches [rectangle] 24 inches — This rectangle →
288 square inches
2 square feet
1800 square centimeters

   Thus we need to look at the numbers and what the numbers mean. For example, suppose you have heard that the floor space of the student center is 10,000 square feet. What does that number mean? Write your thoughts and then share them with your partner(s).

3. Now let us explore the meaning of numbers in measurement. Mikala has a problem:

   The area of her lawn is 1200 square feet. She wants to spread fertilizer over the lawn, but the bag of fertilizer says that one bag will cover 100 square yards. How many bags of fertilizer should she buy?

   a. Think about this problem—what you know about feet and yards, and the tools you have for problem-solving (such as Polya's four steps on the inside front cover of the *Explorations* manual). Write down your initial thoughts and hypotheses, including an estimate if you feel you can make one.

   b. Meet with your partner(s). Select one or more ideas to pursue and then pursue them.

4. After the class discussion, write your second-draft response to the question asked in Step 2. When you see a measurement (for example, 20,000 square feet), what does that number mean?

Mathematics for Elementary School Teachers
p. 621

EXPLORATION 10.7   Exploring Area on Geoboards

Let us explore the concept of area and some area formulas using Geoboards.

PART 1: Area on the Geoboard

In our explorations on the Geoboard and on Geoboard Dot Paper, we will discuss the areas of figures in reference to the unit square—that is, the area enclosed by the smallest square you can make on your Geoboard.

1. **a.** Make as many "different" squares as you can on a 5 × 5 Geoboard or on the Geoboard Dot Paper supplied at the end of the book.

    b. Compare your results and your strategies with your partner(s).

    c. Determine the area of each of the squares.

    d. *Challenge:* Can you make a square with an area of 1 unit? 2 units? 3 units, and so on? Which squares can't be made? Can you explain why they are impossible?

2. **a.** Make as many "different" rectangles as you can on a 5 × 5 Geoboard or on the Geoboard Dot Paper provided by your instructor.

    b. Compare your results and your strategies with your partner(s).

    c. Determine the area of each of the rectangles.

    d. *Challenge:* Can you make a rectangle with an area of 1 unit? 2 units? 3 units, and so on? Which rectangles can't be made? Can you explain why they are impossible?

3. **a.** Determine the area enclosed by each of the figures on page 311. Briefly describe your solution path so that a reader could see how you determined the area.

    b. Compare your answers and your strategies with your partner(s).

    c. With your partner, make an unusual shape on the Geoboard. Determine the area independently, and then compare answers and solution strategies. Do this for several figures.

    d. Summarize and justify useful strategies that you developed. Imagine writing this and sending it to a friend in order to share your insights.

Figures for EXPLORATION 10.7, PART 1: Area on the Geoboard

PART 2: Understanding area formulas

1. ***Determining the area of parallelograms***   Even if you remember certain area formulas, the purpose of these explorations is not to get the formula but to understand why it works. The *why* enables us to apply our understanding to new situations.

    a. Determine the area of the parallelograms below by a means other than using a formula. Can you see any patterns that hold for all four parallelograms? (Make more parallelograms if you wish.)

    b. State and *justify* a formula for determining the area of *any* parallelogram.

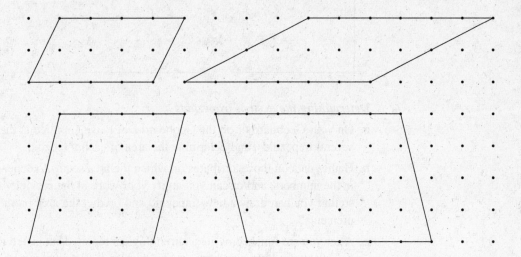

2. ***Determining the area of triangles***   We can apply our understanding of the area of parallelograms to determine a way to find the area of any triangle.

    a. Determine the area of the triangles below by a means other than using a formula. Can you discern any commonalities that enable us to determine the area of any triangle? (Make more triangles if you wish.)

    b. State and *justify* a formula for determining the area of *any* triangle.

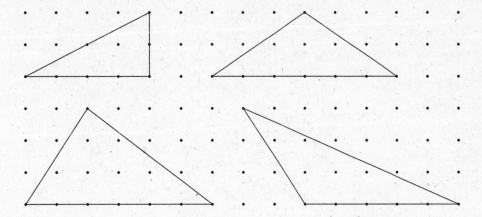

3. There is an old saying that "seeing is believing." Determine the areas of the triangles below—the right triangle, the obtuse triangle, and the isosceles triangle. What did you discover? Can you explain why what you discovered is true?

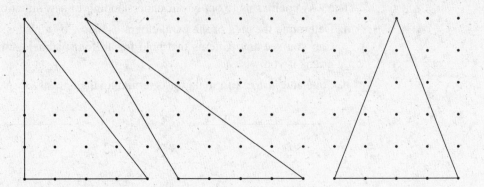

4. *Determining the area of trapezoids*

   a. On your Geoboard or on the Geoboard Dot Paper supplied at the end of the book, make several trapezoids, and determine the area of each trapezoid.

   b. Gather data on those attributes on which the area seems to depend. Can you see patterns in the numbers, and/or can you apply your work in the previous steps in this exploration so that you can draw a new trapezoid and predict the area on the basis of certain measurements?

   c. After you are finished, write a chronological report. That is, tell the reader not only what you discovered but also how you discovered it. If you went down blind alleys or made and then discarded a hypothesis, help the reader to see why the blind alley or hypothesis made sense at the time and what you learned from that work.

*Mathematics
for Elementary
School Teachers*
p. 627

EXPLORATION 10.8   Exploring the Area of a Circle

The purpose of this exploration is to develop an understanding of the formula for the area of a circle.

1.   Cut the circle (at the end of this book) into sectors as shown below. Arrange these sectors into a "parallelogram," as shown.

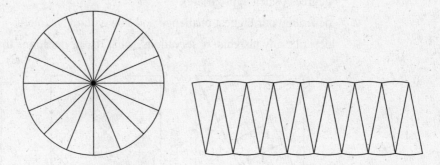

2.   If the radius of the circle is r, determine the length and width of the parallelogram in terms of r.

3.   Use this information to determine the area of the circle. Explain both the *what* and the *why* of your work.

4.   Describe any questions you have about π or circles at this point.

EXPLORATION 10.9   Can You Make the Quilt Pattern?

Each of the following quilt blocks presents different challenges. For each block that you make,

1.   explain any quilt structures (such as patches, an 8-point star, or a hexagon).

2.   justify your choice of dimensions for the whole block (not all are squares).

3.   give step-by-step directions for making the figure. Include the length of each segment and degrees (when appropriate).

4.   describe your biggest challenge and how you overcame it.

5.   describe any insights, observations, what ifs, or questions that you have.

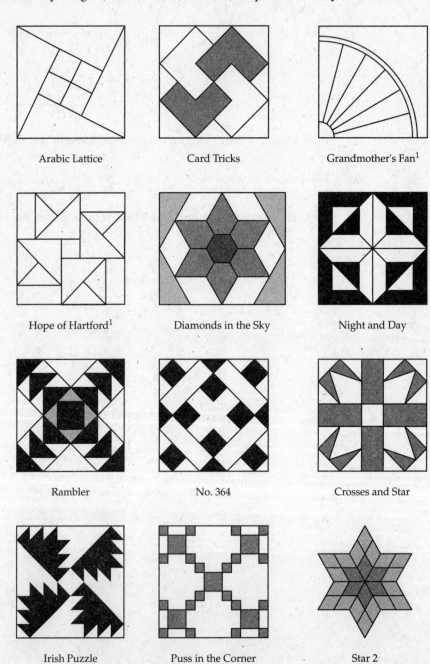

Arabic Lattice   Card Tricks   Grandmother's Fan[1]

Hope of Hartford[1]   Diamonds in the Sky   Night and Day

Rambler   No. 364   Crosses and Star

Irish Puzzle   Puss in the Corner   Star 2

EXPLORATION 10.10 How Much Will the Carpet Cost?

I have seen problems like the following in many textbooks:

> Betty and Bruce have decided that the old carpeting in their living room has to go. They saw an advertisement for carpeting at $15.95 per square yard. Their living room measures 25 feet by 18 feet. How much will the new carpet cost?

The correct "mathematical" answer is obtained by the following process:

1. Find the area of the room: $25 \cdot 18 = 450$ square feet.
2. Convert from square feet to square yards: 450 square feet ÷ 9 square feet/square yard = 50 square yards.
3. Multiply this area by $15.95: (50 square yards)($15.95/square yard) =$797.50.

There is a major problem with this process, however: This is not at all how the carpet store determines how much your carpet will cost.

1. Brainstorm how the carpet store might determine the cost. Imagine that you are the manager of the store and/or that you are the person who installs the carpet. Summarize the results of your brainstorming as statements and/or questions.
2. After the class discussion, summarize the relevant information you will use to determine the cost of the carpet.
3. Now determine how much the carpet store would actually charge Betty and Bruce.

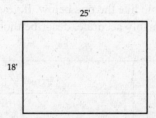

4. Determine the cost of carpet for each of the rooms below.

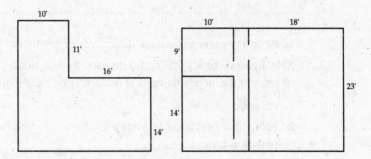

5. Describe any questions you have about carpeting at this point.

*Mathematics
for Elementary
School Teachers*
p. 629

EXPLORATION 10.11   Irregular Areas

Outside the classroom, when we want to find an area, more often than not the object is not one
for which we have a nice, simple formula. However, we can use our knowledge of area to determine the areas of these objects. Applying your understanding of what area means, determine
which of the two ponds below is bigger.

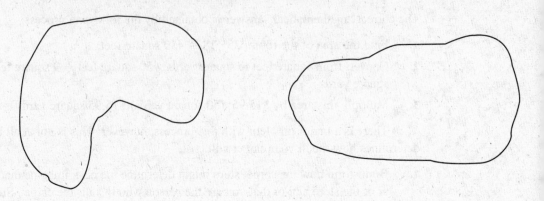

1.   Brainstorm different ways to determine which of these two ponds is bigger (feel free to use
      your imagination). Briefly summarize the methods that arose in your group.

2.   **a.** During the class discussion, make notes of the pros and cons of each method, using a
          table like the one below. If you think any of the methods are not valid or will not be rea-
          sonably accurate, describe those concerns in your "cons."

	Pros	Cons
Method 1		
Method 2		
Etc.		

      b. Note any other important statements or points made in the class discussion.

3.   Select two methods that could be used to determine the actual area, and determine the area
      of each of the ponds using that method. For each method,

      a. describe your work.

      b. justify the precision in your answer—the choice of units, and the number of decimal
          places, if any.

      c. describe any difficulties you had and how you overcame them.

      d. describe any changes or modifications in your ideas with respect to the methods dis-
          cussed in the previous step.

4.   Describe any questions you have about measuring the areas of irregularly shaped objects at
      this point.

*Mathematics
for Elementary
School Teachers*
p. 630

EXPLORATION 10.12   Exploring Relationships Between
Perimeter and Area

The relationship between perimeter and area is a rich field for exploration because there are many possible ways in which these two attributes can be related.

PART 1: Perimeters and areas on Geoboards

1. **a.** Suppose someone doubled the area of her garden. Does that mean that the length of the fence around the perimeter of her garden doubled also? What do you think?

    b. If we know that the area of one garden is 50 percent greater than the area of another garden, does this mean that the perimeter will be 50 percent greater? What do you think?

    c. Take some time to describe your present thoughts on the relationship between perimeter and area from this perspective.

The relationship between area and perimeter is not a simple one, and we are going to investigate the relationship systematically, using a technique developed by mathematicians that helps us when relationships are complicated. We are going to keep one variable constant and look at what happens to the other variable.

2. **a.** On dot paper, make a number of polygons, each of which has an area of 15 units and in each of which all sides are either horizontal or vertical line segments. Determine the perimeter of each figure. In this case, we are holding area and certain attributes of shape constant and looking at how the perimeter changes.

    b. Look at those figures with the smallest perimeters and those with the greatest perimeters. Describe differences between the figures with smaller perimeters and the figures with greater perimeters.

3. **a.** Now make a number of polygons, each of which has a perimeter of 24 units and in each of which all sides are either horizontal or vertical line segments. Determine the area of each figure. In this case, we are holding perimeter and certain attributes of shape constant and looking at how the area changes.

    b. Look at those figures that have the smallest areas and those figures that have the largest areas. Describe differences between the figures with smaller areas and the figures with larger areas.

4. Finally, consider the original question: How are perimeter and area related? What do you believe now? Describe your present beliefs. If they are different from or more refined than your previous beliefs, describe the experiences, observations, or conversations that changed your beliefs.

PART 2: Changing dimensions

Let us explore the relationship between perimeter and area from another perspective.

1. What if you doubled the length of a rectangle but didn't change the width? The area would double, but what about the perimeter? What would be the effect on the perimeter? Explore this question with rectangles of different dimensions. Can you arrive at a statement that will be true for *all* rectangles? Write a report containing the following material:

    a. A brief summary of what you did

    b. Your present description, in words and/or a formula, of the relationship between the original and the new perimeter

2. Exchange descriptions with a partner. Provide feedback with respect to accuracy and clarity.

PART 3: The banquet problem

Spaghetti and Meatballs for All! is a children's book written by Marilyn Burns to engage children in exploring the relationship between area and perimeter. Her story is a variation of a classic math problem called the *banquet problem*. Here we will explore this version: A family is having a get-together. They expect 18 people. They can rent card tables, each of which can seat 4 people. They have decided that they want to arrange the tables in the form of a rectangle. If they simply connect them in one long line, they will need 8 tables, because they can seat 8 people on each side and 1 person on each end. What are all the possibilities?

PART 4: Problems from the classroom

The following problems come from "Perimeter or Area: Which Measure Is It?" by Michaele Chappel and Denisse Thompson, which appears in the September 1999 issue of *Mathematics Teaching in the Middle School*, page 21. The authors report how students did on several questions.

1. Draw a figure whose perimeter is 24 units.

2. Here are two student solutions. Are they correct? Why or why not?

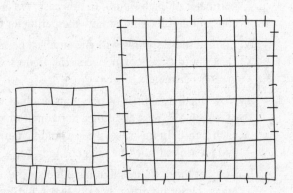

3. Write a realistic story problem in which you need to find the perimeter.

4. Here are three student stories. Discuss the stories in terms of how realistic they are and how valid they are.

   a. Alex has a table that is 100 centimeters long and 50 centimeters wide. What is the perimeter?

   b. Mary wants to border a cake with frosting. The cake is square, and one side is 6 inches. She uses a tablespoon for every inch around the cake. How much frosting does she need?

   c. Ashleigh is buying wallpaper for a room and knows that one wall is 7 feet across and that the floor of the room is a perfect square. Find the perimeter and tell how much wallpaper Ashleigh needs to buy.

5. This question was the hardest. Less than 8% of the students answered the problem correctly. Try it: Can two figures have the same area but different perimeters? Explain your answer.

EXPLORATION 10.13 **Functions, Geometric Figures, and Geoboards**

Throughout this book, we have seen that there are many functional relationships in mathematics and in everyday life. In this exploration, we will discover relationships among three variables: the area of a figure, the number of pegs on the border of the figure, and the number of pegs in the interior of the figure.

PART 1: Exploring the relationships between pegs and right triangles

1. Use your Geoboards or the Geoboard Dot Paper supplied at the back of the book to fill in the top table on page 323. When you see a pattern, note it. Does that pattern lead to a hypothesis? For example, can you predict the area or number of border pegs or interior pegs in the next triangle?

2. When you feel that you are able to predict the area and the number of border and interior pegs when you are given only the dimensions of the triangle, state your hypothesis and how you came to discover it.

3. Predict the values for the three columns of the table for a 16 by 3 triangle and a 20 by 3 rectangle. Explain your predictions.

4. **a.** Use the top grid on page 325 to show the relationship between the length of the longer leg and the area of the triangle.

   b. What does this graph's being a straight line mean? Justify your response.

5. Use the bottom grid on page 325. In this step you are going to graph two different relationships on the same graph.

   a. First plot the points representing the relationship between the length of the longer leg and the number of border pegs. Then plot the points representing the relationship between the length of the longer leg and the number of interior pegs.

   b. What does neither of these graphs' being a straight line mean? Justify your response.

   c. Does it mean that they are not functions? Justify your response.

   d. Can you make use of these patterns to extend the graphs—that is, to predict the number of interior and border pegs when the longer base is 12, 13, 14, 15, or more units long?

   e. State your hypothesis now for predicting the number of interior and border pegs for any right triangle. Justify your hypothesis.

PART 2: Expanding our exploration to other figures

We are going to expand our exploration of relationships among border and interior pegs to include other figures. We are also going to change the focus slightly. Now, we are going to look for patterns so that we can predict the area of any geometric figure on the Geoboard from the number of border and interior pegs.

1. Make several geometric figures on your Geoboard or Geoboard Dot Paper. In each case, determine the area (A), the number of border pegs (B), and the number of interior pegs (I) and record those values in the bottom table on page 323. Do you notice any patterns that lead to hypotheses? If you do, record your hypothesis and how it came to be. Test your hypothesis. If it works, great. If not, back to the drawing board. If you do not see patterns that lead to hypotheses, move on to another geometric figure.

2. Now that you have determined the relationship among the three variables B, I, and A, look back and describe the moment of discovery.

Tables for EXPLORATION 10.13, PART 1 and PART 2

PART 1: Exploring the relationships between pegs and triangles

1.

Dimensions of the right triangle	Area	Number of border pegs	Number of interior pegs	Insights/Observations
3 by 3	4.5 sq. units	9	1	
4 by 3				
5 by 3				
6 by 3				
7 by 3				
8 by 3				
9 by 3				
10 by 3				
11 by 3				

PART 2: Expanding our exploration to other figures

1.

Geometric figure	Area (A)	Number of border pegs (B)	Number of interior pegs (I)

Grids for EXPLORATION 10.13, PART 1: Exploring the relationships between pegs and triangles

4.

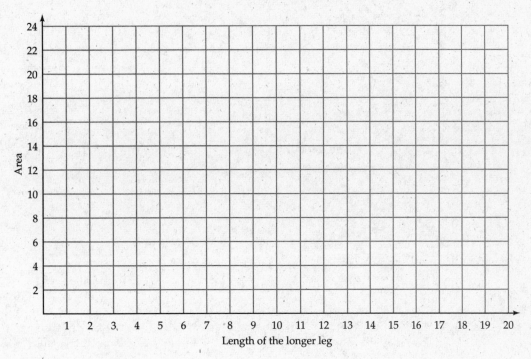

5.

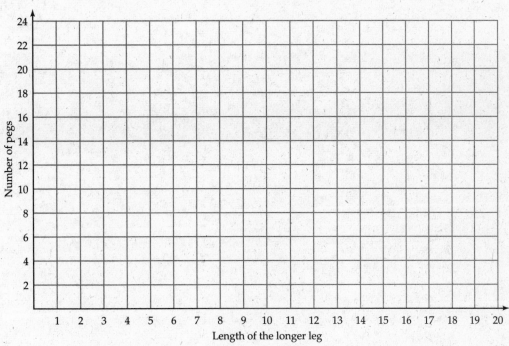

SECTION 10.3   Exploring Surface Area and Volume

Many applications of measurement involve three-dimensional surfaces—either covering them (surface area) or filling them (volume).

Mathematics for Elementary School Teachers
p. 638

EXPLORATION 10.14   Understanding Surface Area

1.   How would you determine the surface area of each of the objects below? Write down your thoughts.

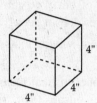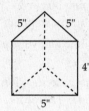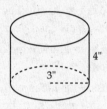

2.   Meet with your partners and pool your ideas. Write down those ideas that the whole group understands and agrees with. The purpose of this step is not for someone who knows the formula to explain it to others so that they can copy it down. The purpose of this step is to discuss ideas and then note those ideas that make sense to the whole group.

3.   Each person in the group will now construct a net (see Exploration 8.19, Part 2: Making Nets) to make one of the models above. Remember that a net is a connected two-dimensional figure that can be folded up to make a space figure. Test each net by folding it to make the shape.

4.   Each person presents a net and describes it to the rest of the group.

5.   Now describe how to find the surface area of each of the figures.

6.   **a.** Describe the commonalities and differences among the first two figures, using a table like the one below.

Commonalities	Differences

b. Describe a general formula for determining the surface area of any *prism*.

Looking Back on Exploration 10.14

Describe any questions you have about surface area.

*Mathematics
for Elementary
School Teachers*
pp. 641, 647

EXPLORATION 10.15 Understanding Volume

PART 1: Volume of a cube

1. You probably know the formula for determining the volume of a rectangular prism: length × width × height. Can you justify this formula—that is, can you explain why we multiply these three numbers? Write your first-draft ideas.

2. Your instructor will give you a rectangular prism to fill with cubes. Do so and then revisit your response to the previous question. If you wish to revise it, do so now, and also explain what happened that led you to decide to revise your answer.

3. A manufacturer sells widgets. Each widget is put in a box whose dimensions are 4 cm × 6 cm × 10 cm. The shipping box is 80 cm × 60 cm × 40 cm. If the company makes 100,000 widgets each year, how many shipping boxes does it need?

   a. Using your knowledge of volume and your problem-solving tools, begin to work on this problem.

   b. Discuss your ideas with your partner(s). Remember to treat ideas as hypotheses: They may or may not be valid. Therefore, the purpose of the group conversation is to convince yourselves of the validity or nonvalidity of hypotheses that are raised.

   c. Describe your solution path and answer.

4. Let's say the company also manufactures gizmos. Suppose the dimensions of the box containing a gizmo are 3.5 cm × 7.6 cm × 11.3 cm. and your shipping box is 80 cm × 60 cm × 40 cm. How should the company arrange the gizmo boxes so that each shipping box will contain the greatest possible number of gizmos?

   a. Using your knowledge of volume and your problem-solving tools, begin to work on this problem.

   b. Discuss your ideas with your partner(s), remembering to treat ideas as hypotheses [as in Step 3(b)].

   c. Describe your solution path and answer.

5. Let's say each object is placed in a box whose dimensions are a, b, and c, and the dimensions of the shipping box are x, y, and z. Tell under what conditions the following formula will work for determining the number of small boxes that one shipping box will contain:

$$\text{Number of objects} = \frac{xyz}{abc}$$

6. The United States Postal Service sells boxes. At the time of the writing of this book, these were the dimensions and prices of the boxes:

8 inches × 10 inches × 12 inches	$1.98
10 inches × 12 inches × 15 inches	$3.19
Cylindrical tube 2 inches (diameter) × 16 inches	$2.19
Cylindrical tube 2 inches (diameter) × 24 inches	$2.49

   a. Are the prices of the boxes proportional to the volumes? Use your knowledge of volume and proportion and your problem-solving tools to work on this problem. State and justify your answer.

   b. Are the prices proportional to the surface areas? State and justify your answer.

   c. Which do you think is the more important measure in determining the cost of the boxes—surface area or volume? Justify your response.

PART 2: Volume of a cylinder

1. Because its base is not a rectangle, we cannot determine its volume by filling it with little cubes. We must use reasoning to develop and then test our conjectures with respect to what its volume might be. Describe your present thoughts concerning how to find the volume of a cylinder. Justify your reasoning.

2. There are several activities that you can do to justify and better understand how to find the volume of a cylinder. Below are several possibilities that will serve as hints. Explore one or more of these hints, and then summarize what you did and what you learned from the exploration.

    a. Imagine making prisms in which the base has more and more sides.

    b. Mark lines on your cylinder at 1 inch high, 2 inches high, and so on. Imagine slicing the cylinder at these points so that your cylinder consists of four of these slices.

3. State the formula for finding the volume of a cylinder and provide your justification of this formula in your own words.

PART 3: Volume of a pyramid

1. Describe your current thoughts concerning how to find the volume of a pyramid. It is important to note that "your current thoughts" is not the same as "your guess as to the formula." Rather, it is your thoughts about how we might find the volume of a pyramid. Your thoughts may contain some quantitative thinking—for example, recall Investigation 10.2D, concerning the area of a circle, where we found that the area of a circle must be less than $4r^2$.

2. If you look closely at a pyramid and a cube, you will find some commonalities. If you haven't noticed them yet, examine these figures and discuss them now in the group. Consider how they might provide more clues about the connections between determining the volume of a cube and determining the volume of a pyramid. Write down your thoughts.

3. Your instructor will provide you with some material. Fill your pyramid and see how much greater the volume of the cube is than that of the pyramid. Before doing so, state your guess and any reasoning. Then do so.

4. Describe your present thinking about the relationship between the volume of a cube and the volume of a pyramid. State your thinking as carefully as possible. For example, it would not be accurate to say that the volume of a cube is greater than the volume of a pyramid; to be accurate, such a statement would have to indicate the conditions under which this would be true.

Looking Back on Exploration 10.15

Describe any questions you have about volume at this point.

Mathematics for Elementary School Teachers p. 647

EXPLORATION 10.16

Determining Volumes of Irregularly Shaped Objects

PART 1: Getting started and making predictions

1. **a.** When you go to the store, which size egg do you buy: medium, large, extra large, or jumbo? Why?

    b. Why don't we call the smallest egg size *small* instead of *medium*?

2. Are you getting the same value for your money with different-sized eggs? That is, would an egg that is twice as big cost twice as much? Write down your initial thoughts.

3. In mathematical terms, what would it mean to say that you are getting the same value when you buy a medium egg as when you buy a jumbo egg?

4. Do you think each of the sizes will be about the same value, or do you think they will vary (remember the 8-inch and 16-inch pizzas from Investigation 10.2E)? Explain your reasoning.

PART 2: Volume and cost

1. **a.** Predict the variation you would expect if we took 100 eggs of the same size and found their volumes. For example, what would be the ratio of the volumes of the largest and smallest eggs? Explain your prediction.

    b. If you were to graph the data with the volume as the independent variable and the frequency of each size as the dependent variable, do you think the graph would be uniform, normal, bimodal, or random? Explain your prediction.

2. Determine the volumes of the eggs and then determine the average volume for each size of egg.

PART 3: Mass and cost

1. **a.** Predict the variation you would expect if we took 100 eggs of the same size and found the mass of each. For example, what would be the ratio of the masses of the largest and smallest eggs? Explain your prediction.

    b. If you were to graph the data with the mass as the independent variable and the frequency of each size as the dependent variable, do you think the graph would be uniform, normal, bimodal, or random? Explain your prediction.

2. Determine the masses of the eggs and then determine the average mass for each size of egg.

PART 4: Mass and volume

1. Do you think the masses and volumes of different eggs will be proportional? First, explain in your own words what the question means. Then write down your hypothesis and explain your reasoning.

2. Using your data from Parts 2 and 3, determine the answer to the question.

Extension

How do you think eggs are actually sized: by mass, volume, length, width, or something else?

EXPLORATION 10.17 Paper Towels

PART 1: Initial questions and ideas

Most households in the United States use paper towels. Putting aside for the moment the issues of conservation and waste, let's say you were in a supermarket and you were deciding which paper towel to buy.

1. What are some of the factors that might enter into your decision to buy a particular brand of paper towel?

2. How would you determine which brand of towel is the "cheapest"?

PART 2: Absorbency

1. Let's say you are more interested in absorbency than in anything else. That is, you want the towel that will absorb the most liquid. How might you determine the most absorbent towel? Write down your initial ideas. After discussing ideas with your partner(s), develop a plan and then determine the most absorbent towel.

2. Your report must contain

   a. your plan and a justification of each step.

   b. your actual data and conclusion. The reader must be able to see what you did and how you arrived at your conclusion.

   c. your justification of the precision in your answer—for example, the choice of unit and the number of decimal places, if any.

   d. a description and explanation of your degree of confidence in your result.

   e. a description of any difficulties you had and how you overcame them, either totally or partially.

3. Present your findings to the class. After the class presentations, describe and explain any changes you would make in your design.

PART 3: Strength

1. Let's say that you are more interested in strength: You want a towel that won't shred when you clean surfaces with it. How might you determine the strongest paper towel? Write down your initial ideas. After discussing ideas with your partners, develop a plan and then determine the strongest towel.

2. Your report must contain

   a. your plan and a justification of each step.

   b. your actual data and conclusion. The reader must be able to see what you did and how you arrived at your conclusion.

   c. your justification of the precision in your answer—for example, the choice of unit and the number of decimal places, if any.

   d. a description and explanation of your degree of confidence in your result.

   e. a description of any difficulties you had and how you overcame them, either totally or partially.

3. Present your findings to the class. After the class presentations, describe and explain any changes you would make in your design.

EXPLORATION 10.18 Measurement, Ambiguity, and Precision

Most textbook problems are well defined, but many, if not most, real-life problems are not well defined. Below are six problems that are not well defined. In some cases, you need to gather more information in order to solve the problem. In all cases, you need to make (and justify) assumptions in order to solve the problem.

Each group will select a question, determine a solution, and present the group solution to the class.

Your report of your findings needs to include

a. your solution, which includes your work and an explanation of your work.

b. your degree of confidence about your answer and a brief explanation of your confidence; if your degree of confidence is not very high, express your uncertainty in the form of questions.

c. a description and justification of the assumptions you made in order to solve this problem.

d. a description of the biggest mathematical problem you encountered along the way and how you overcame it.

e. a description of your one or two most important learnings about mathematics and/or about problem-solving.

Questions

1. McDonald's has now sold over 100 billion hamburgers. Tell us about how much 100 billion hamburgers is. At least one aspect should be to describe how much volume this many hamburgers would take up; you may, but are not required to, represent this answer as a cube.

2. It is estimated that approximately 16 billion disposable diapers are sold each year in the United States. How much volume does this amount of diapers represent?

3. Myra believes that it would be cheaper to make soda cans in the shape of prisms (like children's juice boxes) rather than as the cylinders presently used. Do you agree or disagree with this premise? Describe the pros and cons, both mathematical and other, of making soda cans in the shape of prisms.

4. I heard that Congress had appropriated money to increase office space by 65,000 square feet and to add 18,000 square feet of parking. How many parking spaces can be gotten from 18,000 square feet? Draw a rough blueprint for the parking lot—total dimensions, length and width of parking spaces, the width of the "lanes," and so on.

5. College teachers do not have their own classrooms, so the walls of most college classrooms are rather bare. One day I thought that if we could cover the walls of each college classroom with corkboard, it would make posting things on the wall easier. I wondered how much this would cost. How much would it cost to cover the walls of your present classroom with corkboard?

6. It has been estimated that if we took all the office and writing paper that is thrown away each year, we could build a wall from New York City to Los Angeles that was $8\frac{1}{2}$ inches wide and 7 feet high. If we took all that paper and dumped it onto your campus, how high would that pile be?

EXPLORATION 10.19 **Applying Volume Concepts**

Let's say you started a company that manufactures a healthful, natural cereal. The dimensions of your cereal boxes are 24 cm by 16 cm by 5 cm. However, many customers have asked that you make a bigger box. Design a box that is similar in shape and will hold exactly three times as much cereal.

Your report of your findings needs to include

1. your actual work.

2. your answer.

3. your degree of confidence about your answer and a brief explanation of your confidence; if your degree of confidence is not very high, express your uncertainty in the form of questions.

4. a narrative description of how you arrived at your answer. This description may be a "tour" through your work.

5. a description of the biggest problem you encountered and how you overcame it.

ENDNOTES

CHAPTER 1

p. 13 1. *Introduction to Mathematics* (New York, 1911), pp. 59–69, cited in Robert Moritz, *On Mathematics* (New York: Dover Publications, 1914), p. 199, quoted: A. N. Whitehead.

CHAPTER 2

p. 24 1. *Newsweek,* February 13, 1989, p. 62.

p. 27 2. *Mathematics Teaching in Middle School,* 2(4) (February 1997), pp. 220–224.

CHAPTER 3

p. 67 1. Alfred North Whitehead, *Introduction to Mathematics,* New York, 1911, p. 59, cited in Robert Moritz, *On Mathematics* (New York: Dover Publications, 1914), p. 198.

p. 68 2. Frank J. Swetz, *Capitalism and Arithmetic: The New Math of the 15th Century,* (LaSalle, IL: Open Court Publishing Company, 1987) p. 75.

CHAPTER 5

p. 98 1. In many elementary classrooms, teachers use black chips to represent positive numbers and red chips to represent negative numbers.

p. 110 2. Adapted from Activity 16 in Barbara Reyes et al., *Developing Number Sense in the Middle Grades, Addenda Series, Grades 5–8* (Reston, VA: NCTM, 1991).

p. 110 3. Adapted from Activity 18 in Reyes et al., *Developing Number Sense in the Middle Grades.*

p. 117 4. This problem has been adapted from one developed by Deborah Schifter at Education Development Center.

p. 117 5. This problem has been adapted from a problem in *A Course Guide to Math 010L,* by Ron Narode, Deborah Schifter, and Jack Lochhead, 1985.

p. 119 6. This exploration has benefited from the influence of Ellen Davidson and Jim Hammerman at Education Development Center.

p. 121 7. This exploration is adapted from one developed by Ellen Davidson and Jim Hammerman at Education Development Center.

p. 130 8. George Immerzeel and Melvin Thomas, eds., *Ideas from the Arithmetic Teacher: Grades 6–8* (Reston, VA: NCTM, 1982), p. 45.

CHAPTER 6

p. 149 1. This exploration is adapted from one developed by Nancy Belsky.

p. 150 2. Text adapted from "The 100% Solution," by Scott Kim, *Discovery Magazine,* 1999.

CHAPTER 7

p. 182 1. Stewart Culin, *Games of the North American Indians* (New York: Dover Publications, 1975). Excerpt from *Games of the North American Indians* by Stewart Culin. Copyright © 1975. Reprinted by permission of Dover Publications, Inc.

p. 182 2. Culin, p. 136. Excerpt from *Games of the North American Indians* by Stewart Culin. Copyright © 1975. Reprinted by permission of Dover Publications, Inc.

p. 183 3. Culin, pp. 154–155. Excerpt from *Games of the North American Indians* by Stewart Culin. Copyright © 1975. Reprinted by permission of Dover Publications, Inc.

p. 183 4. Culin, p. 157. Excerpt from *Games of the North American Indians* by Stewart Culin. Copyright © 1975. Reprinted by permission of Dover Publications, Inc.

p. 184 5. Culin, p. 223. Excerpt from *Games of the North American Indians* by Stewart Culin. Copyright © 1975. Reprinted by permission of Dover Publications, Inc.

p. 184 6. Culin, p. 221. Excerpt from *Games of the North American Indians* by Stewart Culin. Copyright © 1975. Reprinted by permission of Dover Publications, Inc.

CHAPTER 8
p. 209 1. Reprinted with permission from *Teaching Children Mathematics,* February 1999, pp. 374–377. Copyright © 1999 by the National Council of Teachers of Mathematics.

CHAPTER 9
p. 264 1. "Which Way Will the Arrow Point?" from Seeing Shapes by Ernest R. Ranucci. In *Geometry and Visualization* by Mathematics Resource Project. © 1977. Creative Publications, Inc.

p. 288 2. Kenneth Boulding, cited in Michael Serra, *Discovering Geometry—An Inductive Approach* (Berkeley, CA: Key Curriculum Press, 1993), p. 88.

p. 288 3. Ann Paul, *Eight Hands Round: A Patchwork Alphabet* (HarperCollins: NY, 1991).

CHAPTER 10
p. 316 1. From Luanne Seymour Cohen, *Quilt Design Masters* (Palo Alto, CA: Dale Seymour Publications, 1996).

INDEX

Cutouts

Base Ten Graph Paper

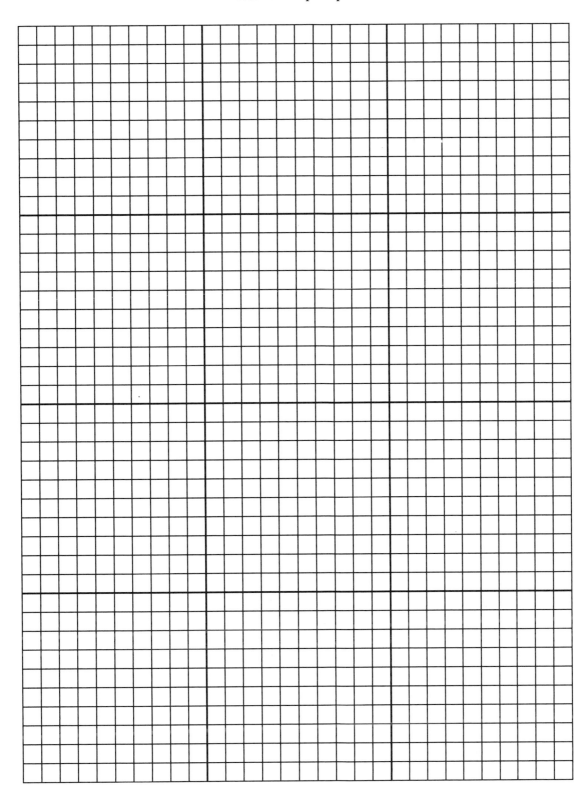

Other Base Graph Paper

Other Base Graph Paper

Other Base Graph Paper

Other Base Graph Paper

Geoboard Dot Paper

Isometric Dot Paper

Polyomino Grid Paper

Polyomino Grid Paper

Polyomino Grid Paper

Polyomino Grid Paper

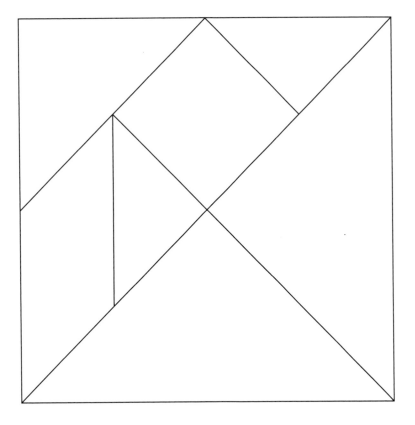

Regular Polygons

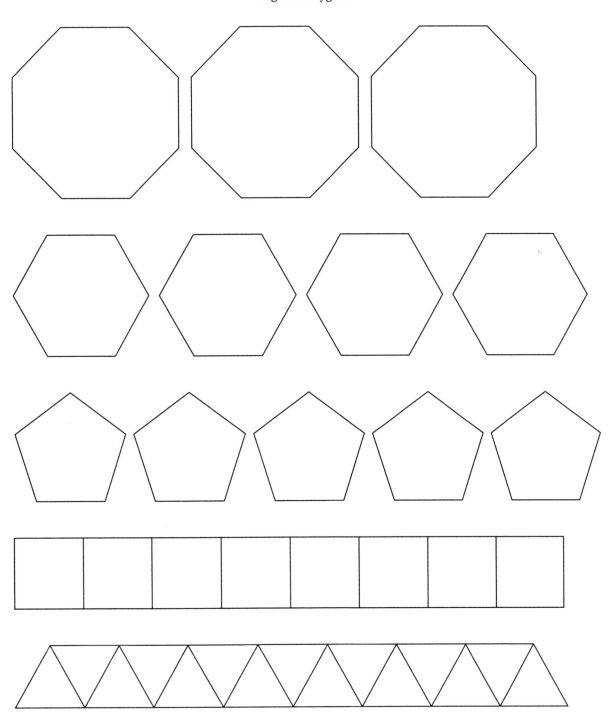

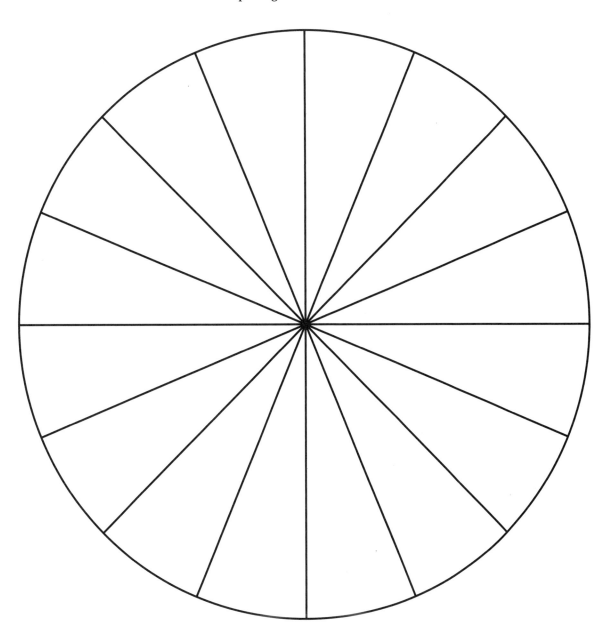